The Rural Text-Book Series

EDITED BY L. H. BAILEY

THE CORN CROPS

The Rural Text-Book Series

LYON AND FIPPIN, PRINCIPLES OF SOIL MANAGEMENT.

G. F. WARREN, ELEMENTS OF AGRICULTURE.

A. R. MANN, BEGINNINGS IN AGRICULTURE.

J. F. DUGGAR, SOUTHERN FIELD CROPS.

B. M. DUGGAR, PLANT PHYSIOLOGY, WITH SPECIAL REFERENCE TO PLANT PRODUCTION.

G. F. WARREN, FARM MANAGEMENT.

M. W. HARPER, ANIMAL HUSBANDRY FOR SCHOOLS.

E. G. MONTGOMERY, THE CORN CROPS.

H. J. WHEELER, MANURES AND FERTILIZERS.

(*Frontispiece*)

TYPICAL PLANTS OF DENT CORN.

THE CORN CROPS

A DISCUSSION OF MAIZE, KAFIRS, AND SORGHUMS AS GROWN IN THE UNITED STATES AND CANADA

BY

E. G. MONTGOMERY

PROFESSOR OF FARM CROPS IN THE NEW YORK STATE COLLEGE OF AGRICULTURE AT CORNELL UNIVERSITY

New York
THE MACMILLAN COMPANY
1916

Set up and electrotyped. Published August, 1913. Reprinted July, 1915; February, 1916.

Norwood Press
J. S. Cushing Co. — Berwick & Smith Co.
Norwood, Mass., U.S.A.

PREFACE

In planning a course of study, the author must needs lay out a working plan. He should know the philosophy of his subject and its relation to other sciences. Field crops like other applied sciences has little pure science of its own, but is rather based on other sciences. The subject is not erected so much as a superstructure on other sciences, but rather moves in a progressive way, between them, abstracting such elements from each as contribute to the art of producing the crop under consideration.

The outline on page vi is an attempt to illustrate the logical order of study and relation of other sciences to the study of Crop Production.

The outline below indicates that a knowledge of all the "earth sciences" is fundamental to a study of crop production, hence a student should have a general course in all these sciences with special emphasis on botany (physiology and ecology) and chemistry.

In regard to a particular crop like maize, this knowledge needs special interpretation and application, which is the function of field crops instruction.

The ability to yield with our ordinary crops is far above the average yield. With maize 200 bushels per acre have been produced under optimum conditions, while the average yield is about 26 bushels. Therefore the study of maize production is principally a study of those factors which serve to hinder full development, and thus limit production,

Outline Plan showing the Relation of the Sciences to the Various Phases of Crop Production

Department giving Work	Character of Work	Basic Science involved
	The Plant	
1. Botany Field crops treats application to special crop	Study of normal plants, their histology, physiology, etc., and normal environmental requirements	Botany
	Survey	
2. Field crops will consider the application of the sciences to the particular crop under consideration	Survey of natural conditions as related to the normal, under which it is proposed to cultivate the plant	Ecology (Botany) Meteorology Geology (Geography)
	Adaptation	
3. (*a*) Field crops (Plant breeding)	(*a*) Adaptation of plant to climate and soil	(*a*) Plant breeding (natural and artificial selection) Ecology
(*b*) Soils	(*b*) Adaptation of soil to plant	(*b*) Chemistry Physics Bacteriology
	Protection	
4. (*a*) Field crops; also farm practice for (*b*) and (*c*)	(*a*) Farm practice in preparing and planting fields in order to protect against weeds, drought, rain, etc.	
(*b*) Botany (the diseases)	(*b*) Against fungous diseases	Botany
(*c*) Entomology (the insects)	(*c*) Against insects	Entomology

and the art of maize production is removing or modifying these limiting factors.

Practically the whole problem is involved in securing a perfect harmony between the plant and its environment.

Environment may be classed as climatic factors and soil factors. Over climate we exercise little or no control. Either the plant must be adapted to suit the climate or its production is limited only to those regions where a natural climate is found to which the plant is suited. The natural precipitation is about the only factor assigned to climate, the effect of which can be modified. Where precipitation is excessive, land can be drained, or where deficient, methods of storing the moisture in soil may be adopted. However, within certain limits there is usually an optimum rainfall which favors the largest production.

Soil environment, however, is subject to modification in a very large degree. If proper elements are present in the soil but in an insoluble state, solvents may be added as decaying organic matter, or air be admitted by tillage and the bacterial flora increased. If the proper mineral elements are not present or present in an unavailable form, these elements may be added to the soil, until a normal state of fertility is produced.

After the conditions of adaptation of both plant and soil have been fulfilled so far as practicable, and seed has been planted in suitable soil, it is then necessary to protect. Protection is the principal reason for cultivation. To facilitate cultivation, systems of planting have been devised, as the distribution of the plants in rows, drills, or checks, in furrows or on the level surface.

Protection against insect enemies and fungous diseases is also an important part of production, and is one of the reasons for the practice of rotations.

A large share of farm practice has to do with modifying the soil environment and protection of the crop.

THE PHILOSOPHY OF CROP PRODUCTION

The art of crop production is based on an application of the sciences, (*a*) to producing a natural condition as perfectly adapted as possible to the needs of some particular crop, or (*b*) the adaptation of the crop to certain natural conditions.

The study of crop production for any large region involves a study of four general phases of the subject, as: 1. The plant, its structure, physiology, and normal requirements. 2. A general survey of the region where it is proposed to cultivate the plant, to note how the natural conditions found correspond to the needs of the plant. 3. The adaptation of the plant on the one hand to natural conditions and adaptation of soil on the other to the needs of the plant. Maximum production is obtained when perfect adaptation is secured. 4. Protection is necessary against other indigenous plants, fungous diseases, and insects.

The treatment of subjects in the text follows practically the above plan. The plan also allows a wider use of the text for different classes of students. The first two divisions are technical and should only be studied by students who have training in the sciences involved. With less advanced students the work may begin with Part III, Adaptation. The third and fourth divisions deal with the more practical phases of production and are written in a more popular style, this double use of the book being in mind.

ACKNOWLEDGMENTS. — For furnishing photographs used in illustrating the text, the author is indebted to Mr. Carleton R. Ball, Mr. C. W. Warburton, and Mr. C. P. Hartley, all of the Bureau of Plant Industry. A large number of photographs secured from the Nebraska Experiment Station have also been used, Professor T. A. Kisselbach furnishing several of these. Also the Portland Cement Co., Deere and Co., Janesville Machine Co., Planet Jr. Co., and Sandwich Manufacturing Co. have furnished illustrative material.

E. G. MONTGOMERY.

ITHACA, N.Y.,
January 1, 1913.

TABLE OF CONTENTS

PART I

CORN

CHAPTER I

SECTION I

THE CORN PLANT

CHAPTER II

CHAPTER III

CHAPTER IV

SECTION II

PRODUCTION AS RELATED TO CLIMATE AND SOILS

CHAPTER V

CHAPTER VI

SECTION III

IMPROVEMENT AND ADAPTATION OF THE CORN PLANT, AND ENVIRONMENT

CHAPTER VII

CHAPTER VIII

CHAPTER IX

CHAPTER X

CHAPTER XI

CHAPTER XII

CHAPTER XIII

CHAPTER XIV

CHAPTER XV

SECTION IV

CULTURAL METHODS

CHAPTER XVI

CHAPTER XVII

PART II

SORGHUMS

CHAPTER XXIII

CHAPTER XXIV

CHAPTER XXV

CHAPTER XXVI

CHAPTER XXVII

CHAPTER XXVIII

CHAPTER XXIX

PART I

CORN

CORN CROPS

CHAPTER I

PRODUCTION AND DISTRIBUTION OF INDIAN CORN

THE corn crops, as understood in this book, are the derivatives of two group-species: of *Zea Mays*, the Indian corn or maize; and of *Andropogon Sorghum*, the sorghum and kafir series. The former is a plant-group of the Western Hemisphere and the latter of the Eastern Hemisphere. The maize products are used both for human and stock food, but the sorghum products are employed in this country mostly for the feeding of animals.

1. Relative importance of corn and other crops in the world. — The hay and forage crop is the most important crop of the world, but this is made up of a great variety of plants. The yield in millions of tons of the world's most important plants is shown in the following diagram: —

WORLD'S CROPS OF THE MOST IMPORTANT FOOD PLANTS. AVERAGE FOR 5 YEARS, 1906–1910

Crop	Millions of Tons
Potatoes	156
Corn	113
Wheat	107
Oats	67
Rice	67
Rye	46
Barley	33

TABLE I

World's Corn Crop by Continents, and Principal Corn-producing Countries. For 5 Years, 1906–1910. (Bushels)

Continent	1906	1907	1908	1909	1910	Annual Average
North America	3,061,470,000	2,755,596,000	2,841,519,000	2,741,448,000	3,095,739,000	2,899,154,400
Europe . . .	608,387,000	489,643,000	529,697,000	535,247,000	644,954,000	561,585,600
South America	198,982,000	78,627,000	143,273,000	185,004,000	183,565,000	157,890,200
Africa . . .	85,544,000	85,402,000	85,426,000	85,807,000	90,846,000	86,605,000
Australia . .	9,262,000	10,912,000	8,907,000	9,644,000	11,863,000	10,117,600
Total . .	3,963,645,000	3,440,180,000	3,608,822,000	3,557,150,000	4,026,967,000	3,715,352,800

Principal Corn-producing Countries

Country	1906	1907	1908	1909	1910	Annual Average
United States .	2,927,416,000	2,592,320,000	2,668,651,000	2,552,190,000	2,886,260,000	2,725,367,400
Austria-Hungary . .	210,472,000	196,620,000	190,651,000	210,684,000	240,761,000	209,837,600
Mexico . . .	110,065,000	140,000,000	150,000,000	170,766,000	190,766,000	152,166,200
Argentina . .	194,910,000	71,768,000	136,055,000	177,155,000	175,187,000	151,015,000
Italy	93,007,000	88,513,000	95,953,000	99,289,000	101,722,000	95,696,800
Roumania . .	130,546,000	57,576,000	78,892,000	70,138,000	103,665,000	88,163,400
Egypt . . .	65,000,000	65,000,000	65,000,000	65,000,000	70,294,000	66,058,800
Russia (Europe)	70,501,000	50,764,000	61,112,000	39,598,000	77,181,000	59,831,200

In total value, the world's wheat crop probably ranks first, the potato crop second, and the corn crop third.

2. Corn crop of the world. — The following tables (I, II) give the world's production of corn for the past five years. The data is abstracted from the Year Books of the United States Department of Agriculture:—

TABLE II

Percentage of World's Corn Crop produced by the Continents, and Principal Corn-producing Countries. For 5 Years, 1906–1910

Continent	1906	1907	1908	1909	1910	Average
North America	77.25	80.56	78.74	77.06	76.88	78.09
Europe	15.34	14.34	14.68	15.04	16.02	15.08
South America	5.03	2.29	3.98	5.20	4.55	4.20
Africa . .	2.15	2.49	2.36	2.42	2.25	2.35
Australia . .	.23	.32	.24	.28	.29	.28
Total . .	100.00	100.00	100.00	100.00	100.00	100.00
Principal Countries						
United States	73.85	75.79	73.94	71.74	71.67	73.39
Austria-Hungary .	5.31	5.74	5.28	5.92	5.97	5.64
Mexico . .	2.77	4.09	4.15	4.77	4.73	4.10
Argentina .	4.91	2.09	3.77	4.98	4.35	4.02
Italy . . .	2.34	2.58	2.65	2.79	2.52	2.57
Roumania .	3.29	1.68	2.18	1.97	2.57	2.34
Egypt . . .	1.64	1.90	1.80	1.82	1.74	1.78
Russia (European)	1.77	1.48	1.69	1.11	1.91	1.59
Total . .	95.88	95.35	95.46	95.10	95.46	95.43

The world's corn crop varies from about three and one-half billion bushels to about four billion bushels, or a variation of 12 per cent. This rather wide variation is due to the fact that more than one-half the world's corn crop is concentrated in one section of the United States.

The comparative production is brought out more clearly in Table II, based on percentage production.

From the tables, it appears that North America produces 78 per cent of the world's corn crop, Europe produces 15 per cent, leaving only 7 per cent for the other continents. The United States produces about 73 per cent of the world's crop, Austria-Hungary 5.6 per cent, Mexico 4.1 per cent, and Argentina 4 per cent, the four countries combined producing 87 per cent of the world's crop.

TABLE III

SHOWING CORN EXPORTED BY COUNTRIES AND PERCENTAGE OF TOTAL WORLD'S EXPORTS FOR 5 YEARS, 1906–1910, INCLUSIVE

COUNTRY	AVERAGE ANNUAL EXPORT BUSHELS	PERCENTAGE OF TOTAL EXPORTS
Argentina	83,569,388	35.66
United States	62,596,444.2	26.64
Roumania	33,124,210.4	14.15
Russia (European)	23,255,489.2	9.90
Belgium	7,007,737.8	2.93
Netherlands	6,718,712	2.83
Bulgaria	6,021,984.4	2.55
Servia	3,054,136.2	1.35
Austria-Hungary	328,352.6	.14
Uruguay	210,674.2	.09
Other Countries	8,703,035	3.75
Total	234,590,164.0	100.00

3. International trade in corn. — The net exports and imports indicate those countries producing a surplus, and those countries as well that must buy. Table III shows that Argentina furnishes about 35 per cent of the world's export corn and the United States only 26 per cent. Table IV shows that Argentina exports 55 per cent of the crop produced, while the United States exports only 2.29 per cent. This country can hardly be classed as a surplus corn country, though the small percentage exported furnishes one-fourth of the world's export corn. The principal importing country is the United Kingdom, taking 36 per cent of the world's trade in corn, and Germany 14 per cent more, the two taking one-half the corn trade.

TABLE IV

SHOWING PERCENTAGE OF TOTAL CORN CROP EXPORTED BY THE PRINCIPAL EXPORTING COUNTRIES, 5-YEAR AVERAGE, 1906–1910, INCLUSIVE

Country	Production in Bushels	Exportation in Bushels	Percentage of Crop
United States . .	2,725,367,400	62,596,444	2.29
Argentina . . .	151,015,000	83,569,388	55.33
European Russia .	59,831,200	23,255,489	38.86
Roumania . . .	88,163,400	33,124,210	37.57
Bulgaria	22,281,800	6,021,984	27.02

Europe consumes about 91 per cent of the world's corn trade. This corn is largely used for feeding live-stock, but also in the brewing industry.

Exportation of corn from the United States is decreasing. The maximum exportation from this country was during the 5-year period 1896–1900, when it reached an annual average of 9.4 per cent. The present decrease in exportation, indicates that home consumption in the United States will soon equal production. In fact, in the past three years corn has been imported on the Pacific Coast.

TABLE V

SHOWING CORN IMPORTED BY COUNTRIES AND PERCENTAGE OF TOTAL WORLD'S IMPORTS FOR 5 YEARS, 1906–1910, INCLUSIVE

COUNTRY	AV. ANNUAL IMPORT OF CORN IN BU. FOR 5 YR.	PERCENTAGE OF TOTAL IMPORT
United Kingdom . . .	84,835,078	36.07
Germany	34,189,007	14.53
Netherland	24,836,943.4	10.56
Belgium	21,984,982.6	9.36
France	13,510,287.2	5.74
Denmark	12,705,123.8	5.45
Canada	10,809,151.8	4.59
Italy	7,737,137.8	3.29
Spain	4,891,501	2.08
Austria-Hungary . . .	4,170,578.2	1.77
Switzerland	2,996,767.6	1.27
Mexico	2,738,086.8	1.16
Cuba	2,546,576.8	1.08
Portugal	1,169,913.4	.49
Norway	1,043,998	.44
Egypt	662,416.4	.28
Sweden	386,611	.16
Russia	329,755.6	.14
British South Africa . .	147,452.2	.06
Other Countries . . .	3,453,661.4	1.46
Total	235,145,030.0	100.00

CORN PRODUCTION IN THE UNITED STATES

4. Relative value of different crops in the United States. — The corn crop is more valuable than any two other crops in the United States. The value of all wealth produced on farms, including that derived from cereals, hay, cotton, live-stock, forests, and fruit, amounts to 7955 millions of dollars. The corn crop alone furnishes about one-fifth of this annual wealth.

RELATIVE FARM VALUE OF PRINCIPAL CROPS IN THE UNITED STATES. AVERAGE FOR 5 YEARS, 1906–1910

5. Development of corn production in United States is shown in the following table: —

TABLE VI

AVERAGE PRODUCTION OF CORN AT DIFFERENT PERIODS

YEARS	ACRES	BUSHELS (000 OMITTED)	YIELD PER ACRE BUSHELS	TOTAL VALUE (000 OMITTED)	VALUE PER BUSHEL
				Dollars	Cents
1849		592,071			
1859		838,793			
1867–1876	38,688,449	1,011,535	26.2	457,000	46.5
1877–1886	68,408,900	1,575,626	25.1	625,623	40.3
1887–1896	74,290,879	1,800,271	24.0	633,694	36.6
1897–1906	87,971,235	2,240,363	25.4	869,575	39.0

The total crop has about doubled in 40 years and quadrupled in 60 years.

6. Production by states. — Table VII gives the most important data summarized on the production of corn by states. This table is arranged according to rank by states and shows that the eight leading states produced about 63 per cent of the total crop.

TABLE VII

ACREAGE, PRODUCTION, AND VALUE OF CORN BY STATES

Average for 1906–1910, Inclusive — Ten-year Average, 1900–1909, Inclusive

State	Production	Acreage	Value	Average Yield per Acre Bushels	Average Price per Bushel	Per Cent shipped out of County where grown
					Cts.	
Illinois	355,907,917	9,776,777	160,243,010	34.5	43	41
Iowa	323,292,400	9,275,400	135,266,400	32.3	39	20
Missouri	222,421,700	7,401,200	107,606,710	28.6	45	12
Nebraska	201,315,100	7,421,800	81,986,185	27.4	38	38
Indiana	174,940,953	4,716,756	79,616,951	34.7	43	30
Kansas	166,223,800	7,585,800	75,326,000	22.4	42	21
Texas	145,764,156	6,823,531	87,194,878	19.0	59	8
Ohio	139,602,600	3,630,200	71,442,910	35.6	48	23
Oklahoma	111,155,518	5,052,917	50,494,408	24.2	44	22
Kentucky	93,633,675	3,359,414	50,842,339	26.7	51	10
Tennessee	80,722,800	3,197,152	47,030,317	23.0	55	14
Michigan	55,840,000	1,707,000	31,017,400	32.7	51	7
South Dakota	55,144,700	1,961,000	23,506,525	27.4	39	23
Minnesota	55,038,855	1,753,317	25,580,150	29.4	41	12
Pennsylvania	52,256,247	1,423,159	32,895,664	34.4	59	6
Wisconsin	51,006,146	1,473,975	27,161,670	33.2	48	5
Georgia	50,942,319	4,006,576	39,419,323	11.5	73	4
Arkansas	49,047,513	2,420,879	30,072,041	18.7	58	4
Virginia	45,907,704	1,889,122	30,134,137	22.7	59	10
Alabama	44,115,278	2,884,917	32,954,522	13.5	69	3
North Carolina	44,078,769	2,671,964	33,564,971	14.8	68	4

Mississippi	42,131,641	2,423,564	30,256,883	15.2	67	2
Louisiana	31,236,126	1,641,856	20,027,318	17.5	64	4
South Carolina . .	27,019,846	1,851,069	22,352,840	11.6	76	3
Maryland	21,788,565	651,959	12,336,304	32.9	52	29
West Virginia . . .	20,657,200	730,800	14,263,150	27.5	63	6
New York	20,271,800	582,400	14,072,030	31.0	66	2
New Jersey	9,803,257	273,149	6,203,522	34.1	59	13
Florida	7,138,000	621,800	5,613,500	10.2	74	3
Delaware	5,683,832	193,294	2,983,109	29.1	50	38
Colorado	4,027,827	205,031	2,534,513	21.2	59	7
North Dakota . . .	3,797,400	172,200	2,042,260	23.4	46	2
Connecticut . . .	2,405,960	55,919	1,719,256	36.8	70	1
Vermont	2,018,286	52,098	1,423,640	34.5	68	0
Massachusetts . . .	1,851,504	44,159	1,361,222	36.1	71	0
California	1,715,962	52,631	1,392,905	31.4	76	14
New Mexico . . .	4,027,827	205,031	2,534,513	21.2	59	7
New Hampshire . .	982,755	24,447	713,723	33.1	70	0
Maine	579,790	14,070	435,089	35.4	72	0
Oregon	459,018	17,016	344,281	25.8	66	2
Washington . . .	458,277	18,088	341,922	23.8	65	3
Rhode Island . . .	373,872	10,002	313,014	33.0	79	1
Arizona	347,825	11,892	349,422	27.1	94	4
Utah	268,206	9,425	204,892	26.9	74	3
Idaho	222,007	7,046	155,180	28.5	65	3
Montana	183,826	7,596	158,507	23.2	73	1
Wyomng	102,651	5,705	73,454	28.0	66	0
Nevada [1]	27,000	1,000	25,000			

[1] Average for 1909-1911, inclusive.

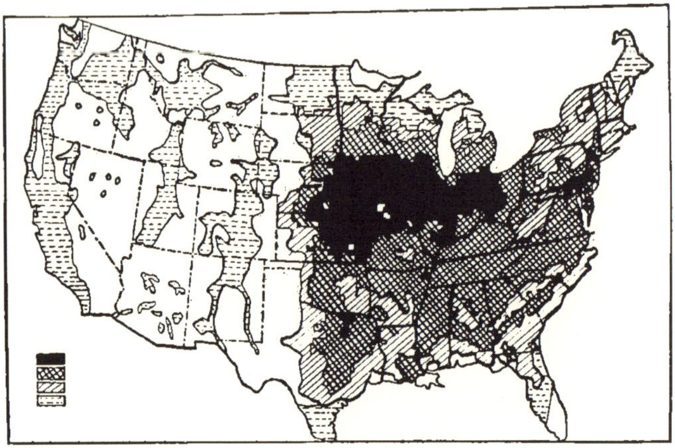

FIG. 1. — Corn production in the United States.

Corn per square mile, census 1900: Black shading, more than 3200 bu.; next shading, 640 to 3200 bu.; next-to-bottom shading, 64 to 640 bu.; bottom shading, less than 64 bu.

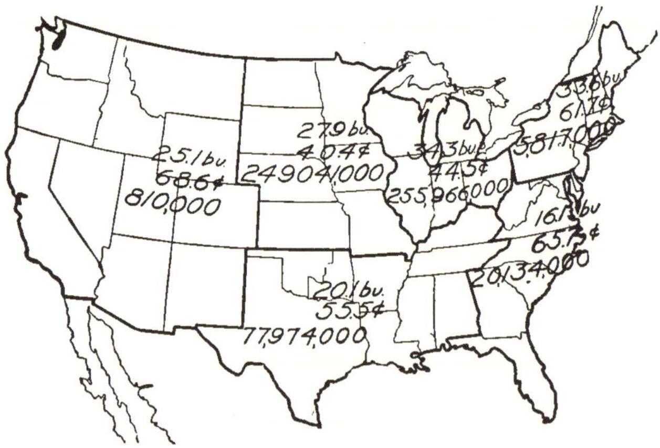

FIG. 2. — Map showing average yield per acre, average farm price per bushel, and average shipment out of county where grown for grand divisions of the United States.

7. Production by sections, and market movement.— The following summary, together with Fig. 2, gives a definite idea of the relative production in different sections of the country, and also the comparative market movement. The available data is corn shipped out of the county where grown, and does not always mean that the corn leaves the state, but indicates the surplus corn in the hands of growers. Most of the lesser corn states consume more corn than they raise, while in the principal corn-belt, most of the corn put on the market leaves the state, and is utilized in manufacturing corn products or shipped to other regions: —

TABLE VIII

TABLE SHOWING PERCENTAGE OF ENTIRE CORN CROP PRODUCED BY EACH GRAND DIVISION OF THE UNITED STATES AND THE MARKET MOVEMENT. 5-YEAR AVERAGE, 1906–1910, INCLUSIVE

Grand Division	Total Production	Per Cent of Total Production Shipped	Amount Shipped	Per Cent of Total Shipped	Per Cent Total Crop of Each G. Div. Shipped
North Atlantic	90,543,473	3.32	5,817,676	.96	6.42
South Atlantic	223,216,236	8.19	20,134,990	3.30	9.02
North Central East Miss. River . .	777,297,616	28.53	255,966,226	41.98	32.93
North Central West Miss. River . .	1,027,233,955	37.69	249,041,882	40.84	24.24
South Central	597,806,892	21.93	77,974,324	12.78	13.04
Far West . .	9,329,243	.34	810,453	.14	8.68
Total . .	2,725,427,415	100.	609,745,551	100.	

SECTION I

THE CORN PLANT

CHAPTER II

ORIGIN AND CLASSIFICATION

In common with all living organisms, corn has been developed through a long and slow evolutionary process. We can only guess at the probable place, origin, and course of evolution by a study of botanically related forms, and especially by a consideration of the embryonic development of the corn plant itself. How much of the evolutionary change was wrought by natural selection, and how much is the result of artificial selection, we can never know. It is probable that corn reached a stage of economic value before attracting the attention or care of man. Since then, no doubt most of the further changes are the result of natural variation and artificial selection.

8. Geographical origin. — Numbers of investigators have made careful studies regarding the probable region in which Indian corn originated. In the early part of the nineteenth century, there was some controversy as to whether this plant was of American origin, the question being based on the contention of some persons that maize had been cultivated in Europe previous to the discovery of America. Careful investigation has not disclosed proof of this supposition, and it is not likely that a plant of such easy culture and obvious value could have existed in Europe without being known. According to Harshberger, it seems most probable that the cultivation of maize originated in the high plateau region of central or southern Mexico at an elevation of about 4500 feet. In this region, plants of *Zea canina* are found growing wild; it is also the native habitat of teosinte and gama grass, two plants closely related botanically to maize. Harshberger concludes that maize probably came into cultivation in this region about the beginning of the Christian Era and spread rapidly both north and south, reaching the Rio Grande about 700 A.D., and the coast of Maine not later than the year 1000.

When Columbus visited America in 1492, maize was in common cultivation. It was at once introduced into other parts of the world, reaching Europe, Africa, China, and Asia Minor early in the sixteenth century. Its early culture in the Eastern Hemisphere seems to have been confined mostly to the countries bordering on the Mediterranean Sea.

Maize acquired many names in Europe, such as Spanish corn, Roman corn, Guinea corn, Turkish wheat, Egyptian corn; these names probably indicate the places where its culture first became extensive.

9. Biological origin. — The Gramineæ, or grass family, includes most of our common cereals, as maize, oats, wheat, and rye. A distinguishing feature of the tribe *Maydeæ*, to which maize belongs, is the separation of its staminate flowers (pollen-bearing) from its pistillate flowers (seed-bearing). Two grasses related to maize and of common occurrence in Mexico — the region in which corn is supposed to have originated — are gama grass (*Tripsacum dactyloides*) and teosinte (*Euchlæna Mexicana*).

Gama grass is distributed also over the southern half of the United States and usually is found on low, rich soil. At a distance a patch of this grass looks very much

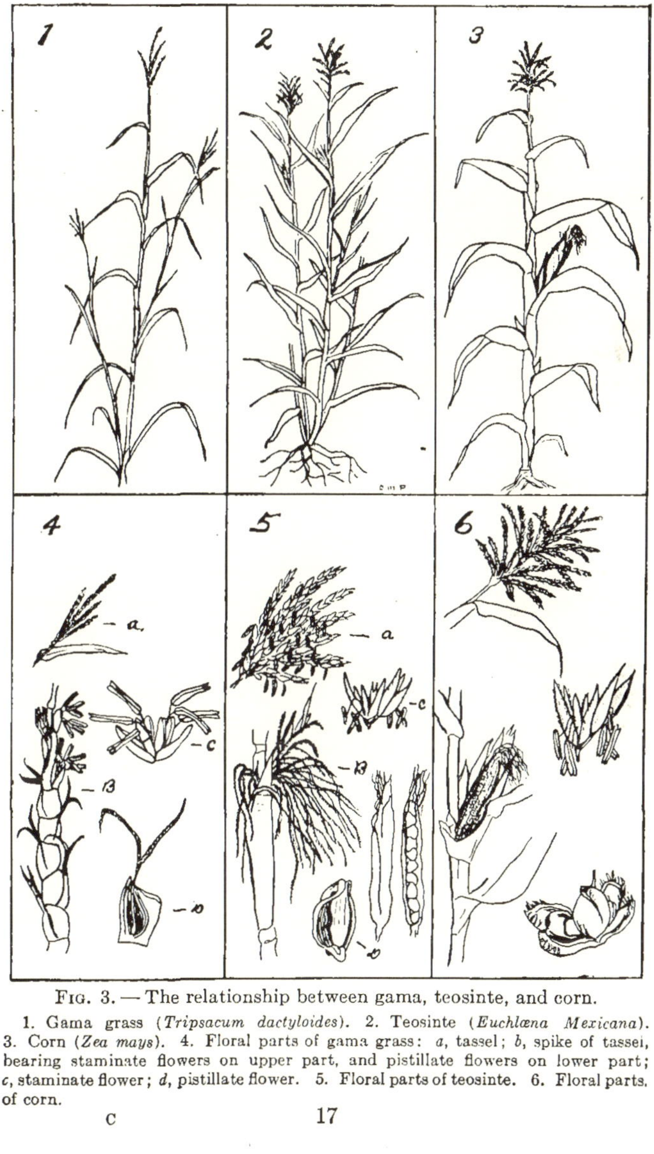

FIG. 3. — The relationship between gama, teosinte, and corn.

1. Gama grass (*Tripsacum dactyloides*). 2. Teosinte (*Euchlæna Mexicana*). 3. Corn (*Zea mays*). 4. Floral parts of gama grass: *a*, tassel; *b*, spike of tassel, bearing staminate flowers on upper part, and pistillate flowers on lower part; *c*, staminate flower; *d*, pistillate flower. 5. Floral parts of teosinte. 6. Floral parts of corn.

like maize. While it grows to a height of five to ten feet, the stem is slender and the leaf about half the width of the maize leaf. The plant bears a tassel-like structure at the top and on the lateral branches, closely resembling the maize tassel, except that the seeds are borne on the lower part of each tassel and the pollen on the upper part.

Teosinte, which is sometimes cultivated but does not mature north of Mexico, is more like maize than is gama grass, the plant being larger and the terminal tassel bearing pollen only. The lateral branches of the plant are so shortened that the terminal tassel-like structure is borne in a leaf axil, surrounded by a kind of husk as is an ear of maize, and bears only pistillate flowers, or seed. It is only a step in the production of an ear of maize, from teosinte, by a development of the central spike of the lateral tassel into an ear.

It is probable that the early progenitor of maize was a grass-like plant having a tassel at the top and tassel-like structures on long, lateral branches, all tassels bearing perfect flowers. As evolution progressed, the terminal tassel came to produce only pollen, and the side branches only ovules, or seeds. Evolution often results in a greater "division of labor," as in this case. At the same time, the lateral branches were shortened or telescoped into the leaf sheaths, these sheaths forming a covering, or husk, for the ear. Also it is probable that in this evolution the central spike of the tassel developed into an ear.

The close relationship of maize and teosinte is proved by the crosses that have been made between the two. In the third or fourth generation after crossing, a peculiar type of corn is secured, identical with a type of maize that has been found growing wild in Mexico (*Zea canina*), and

FIG. 4. — Teosinte and Hybrids with Indian Corn. First two on left are ears of teosinte. The two ears in middle section are hybrids of corn and teosinte. The two on right are fourth-generation hybrids, and are exactly like *Zea canina*. (Grown by U.S. Dept. Agr. near Washington, D.C.)

is supposed by some persons to be the true wild maize and the progenitor of our cutivated maize.

Watson and Bailey both studied this wild maize and regarded it as a distinct species; however, since it has been produced by hybridizing teosinte and maize, this probably accounts for its origin.

CLASSIFICATION OF MAIZE IN GROUPS

Order — *Gramineæ*
Tribe — *Maydeæ*
Genus — *Zea*
Species — *Mays*

10. Maize may be classified into the following groups, or "agricultural species" (after Sturtevant): —

1. *Zea Mays canina* (Watson), Maiz de Coyote. Said to grow wild in Mexico, but the same type has been produced artificially by crossing teosinte and common maize. Characterized by a branching plant and by the production of numerous small ears in the leaf axils of lateral branches; ears sometimes clustered; 4 to 8 rows on an ear, and ear 2 to 4 inches in length.

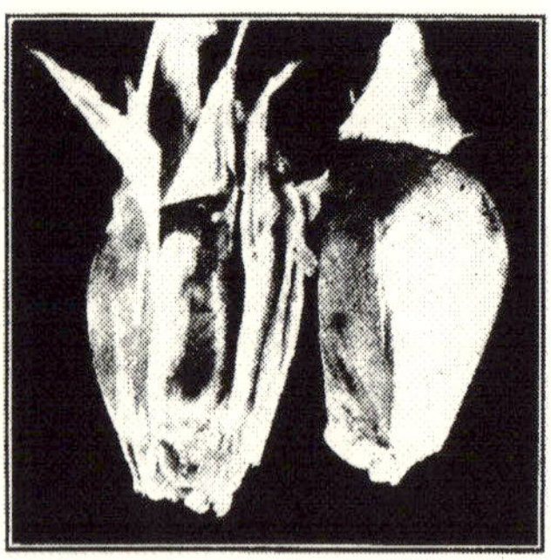

FIG. 5. — Pod corn.

2. *Zea Mays tunicata*, the pod corns, Bul. Torrey Bot. Club, 1904.

Each kernel inclosed in a pod or husk and the ear inclosed in husks; not common. All forms of kernel, as sweet, dent, flint, and others, are found in pod corn. Occasionally a few podded kernels will occur on ears of ordinary corn. It has been supposed by some persons that pod corn represented a primitive or early type of corn, but there is no good evidence for this surmise.

3. *Zea Mays everta*, the pop corns.

Characterized by the excessive proportion of corneous endosperm and the small size of the kernels and ear. The popping quality is due to the explosion of contained moisture on the application of heat, and the best varieties for popping are usually corneous throughout. Two forms of seed are common, one of which is pointed at the top (rice pop corn), and the other form is rounded (pearl pop corn), much as a small flint. All maize colors are found, as red, yellow, white, and blue. The ears are small but vary in length from 2 inches in Tom Thumb to 5 inches for rice and 7 inches for some of the large pearl types. Rows vary from 8 to 16.

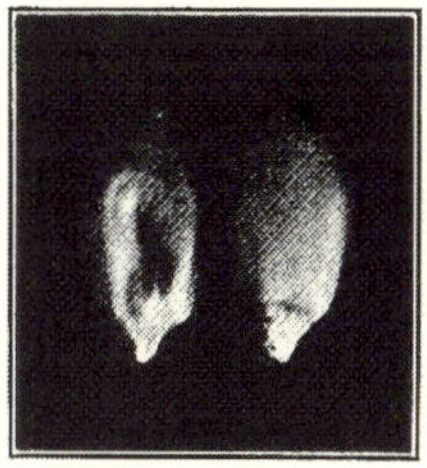

Fig. 6. — Pop corn.

4. *Zea Mays indurata*, the flint corns.

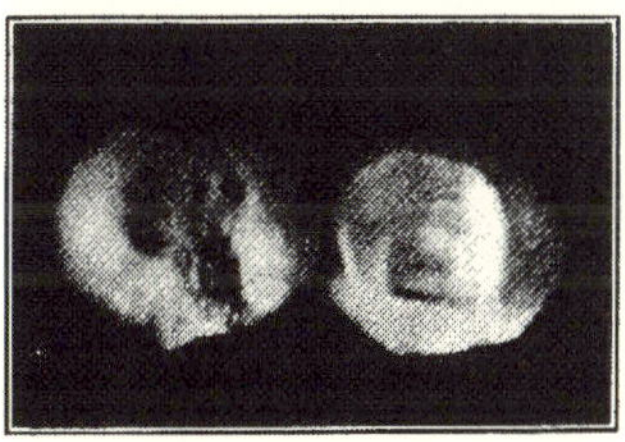

Fig 7. — Flint corn.

Characterized by white starchy endosperm, inclosed by flinty endosperm. Kernels oval in form; in some varieties the corneous part is very thin at the top and a slight indentation appears. There are types of flint maize closely resembling pop corn on the one hand and approaching dent on the other, thus forming a series between the pop and dent corns. Flint maize has all the common maize colors. It varies in length of ear from 8 to

14 inches, and has 6 to 12 rows. The maize most commonly cultivated by the early colonists and North American Indians is extensively cultivated at present in regions where the large dents do not mature.

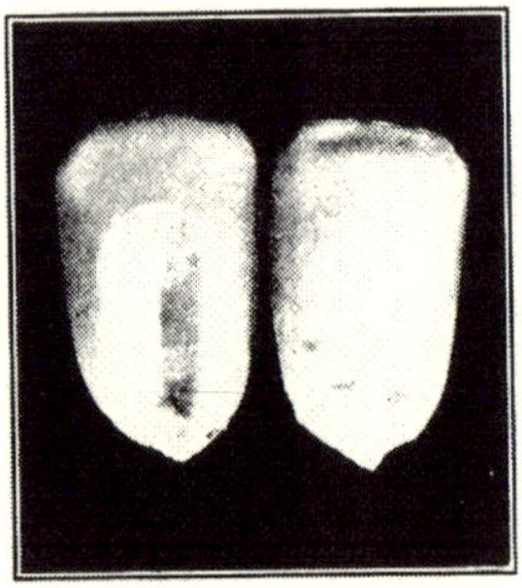

Fig. 8. — Dent corn.

5. *Zea Mays indentata*, the dent corns.

Characterized by horny endosperm at the sides, with starchy endosperm extending to the summit. By shrinkage of the starchy matter in drying, the summit of the kernel is drawn in and indented in various forms. The plant varies in height from 5 to 18 feet; the ear varies in length from 6 to 12 inches and has 8 to 24 rows. The most commonly cultivated type in the United States.

6. *Zea Mays amylacea*, the soft corns.

Characterized by entire absence of corneous endosperm. All soft. No indentations, the kernel being shaped like that of flint corn. Ears mostly 8 to 12-rowed, 8 to 10 inches in length. The usual colors occur. Cultivated to some extent in Southwestern States, Mexico, and South America.

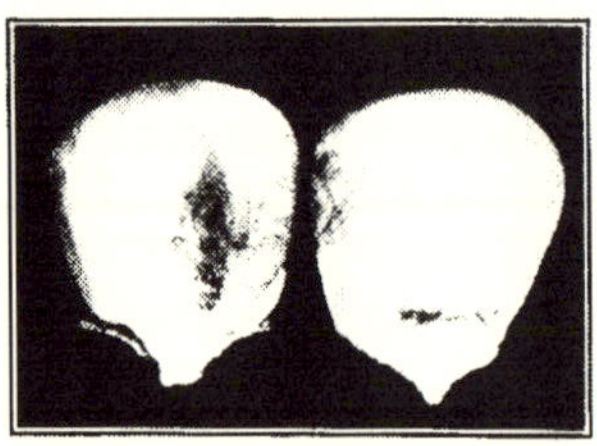

Fig. 9. — Soft corn.

7. *Zea Mays saccharata*, the sweet corns.

Characterized by the translucent, horny appearance and more or less wrinkled condition of the kernel. Shrinking probably due to the conversion of starch into glucose.

According to East, sweet corns are either dent or flint corns that have failed to convert their sugars into starch. Usual variations in color, size, and time of maturity.

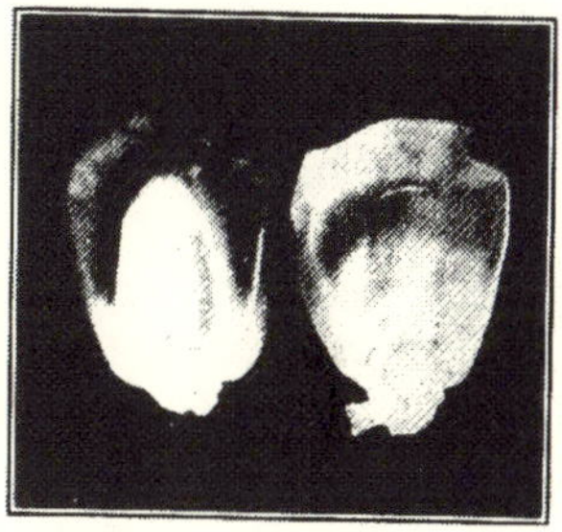

Fig. 10. — Sweet corn.

Zea Mays japonica. The leaves of this species are striped green and white; the grain resembles a pop or small flint type. Cultivated as an ornamental.

Zea Mays hirta. Characterized by an unusual amount of hairs on leaves and sheath, sufficient to be distinctly noticeable. Flint, pop, and dent types. Found mostly in South America.

Fig. 11. — The six principal types of corn. From left to right, pod corn, pop corn, flint corn, dent corn, soft corn, and sweet corn.

Zea Mays curagua. Characterized by a serrate leaf edge. Probably a flint type.

Chinese maize. A small-eared type resembling pearl pop corn, but characterized by a softer, opaque endosperm. Not starchy. A tendency for the upper leaves to be on one side of the plant. (See Bur. Plant Indus., Bul. 161.)

Hermaphrodite forms (perfect flowers). A hermaphrodite form has been described several times. Each pistillate flower bears 3 stamens. The plant is usually short-jointed, with very broad leaves. (See Exp. Sta. Rec., Vol. 18, p. 732. Pop. Sci. Mo. Jan. 1906.; Oct. 1911.)

References on early history: —

DARWIN, CHAS. (1874.) Animals and Plants under Domestication, p. 338.

DE CANDOLLE, A. (1882.) Origin of Cultivated Plants, p. 387.

STURTEVANT, E. L. (1899.) U. S. Dept. Agr., Office of Exp. Sta., Bul. 57.

HARSHBERGER, JOHN W. (1893.) Maize: A Botanical and Economic Study. Bot. Lab. Univ. Penn., Vol. I, No. 2.

COLLINS, G. N. (1909.) U. S. Dept. Agr., Bur. Plant Indus., Bul. 161.

References on biological origin of maize: —

HACKEL. (1890.) True Grasses. Translated by Scribner and Southworth, pp. 36–43.

Grasses of Iowa, Bul. Iowa Geol. Survey, 1903.

HARSHBERGER, J. W. Maize: A Botanical and Economic Study. Bot. Lab. Univ. Penn., Vol. I, No. 2, p. 94.

MONTGOMERY, E. G. (1906.) What is an Ear of Corn? Pop. Sci. Mo., Jan. 1906. Perfect Flowers in Maize. Same, Oct. 1911.

References on *Zea canina*: —

WATSON. Proc. Amer. Acad. Arts and Sci., *26*: 160. Grasses of Iowa. Bul. Iowa Geol. Survey, *1903*: 11–19.

BAILEY, L. H. (1892.) Cornell Univ. Agr. Exp. Sta., Bul. 49.

References to crosses of maize and teosinte: —

HARSHBERGER, J. W. Crosses of Teosinte and Maize. Garden and Forest, *IX*: 522.

U. S. Dept. Agr. Year Book, *1909*: 312.

References on classification of maize: —

BONAFAUS. Mais. folio. Paris, 1836 (folio).

Index Kewensis.

STURTEVANT, E. L. (1899.) Varieties of Corn. Bul. 57, Office of Exp. Sta., U. S. Dept. Agr.

CHAPTER III

DESCRIPTION OF THE CORN PLANT

UNDER the head "Biological Origin" (page 15) it is seen that corn, through a process of evolution, probably came from some branched, grass-like plant resembling teosinte. In Fig. 13 is shown a drawing of a corn plant, with leaves removed, illustrating the grass-like character.

The main stem is divided by *nodes*. Below the ground, the nodes are very close together and give rise to roots; at the surface they give rise to branches or *tillers* and also roots, and above ground to leaves and ears.

The branches or tillers correspond in detail to the main stem, having in all cases as many nodes and leaves as the main stem above the point of attachment. The ear is only a modified branch, as the ear stem has exactly the same number of nodes as the main stem above, and the ear corresponds in many details to the tassel.

11. The root. — When a kernel of maize germinates there is produced, first, a root from the tip end of the seed. A few hours later the stem will appear at the upper end of the germ chit. At nearly the same time two roots will be sprouting from about the median point between root and stem. These are the "temporary" roots and maintain the plant for only a short time. When the corn plant is about six to ten days old, whorls of permanent roots begin to develop at a point about one inch below the ground surface. The seed may be planted 1 to 5 inches deep,

FIG. 12. — Corn roots. 1. Ordinary distribution of roots when corn is planted in rows three feet six inches apart in a deep loam soil. Figures in margin indicate feet. In a hardpan soil roots do not penetrate so deep. 2. Single lateral root. 3. Small branch root showing root-hairs. 4. Root and root-hairs enlarged. 5. Cross-section of 4 at point *a*. 7. Root-hair in contact with soil grains.

but the permanent roots develop at about the same distance below the surface.

12. The spread of the roots. — Root studies on maize at the Wisconsin, Minnesota, Colorado, New York, and North Dakota experiment stations indicate that the permanent roots first spread laterally for about nine to twelve days, when they will have reached a distance 16 to 18 inches from the plant and will be confined mostly to a zone between 3 and 6 inches below the surface. From this time on, the root system rapidly extends downward as well as laterally, at eighteen days reaching a depth of about 12 inches and at twenty-seven days a depth of 18 inches, with a lateral extension of 24 inches. By the time the maize plants are two months old, when they are 5 to 6 feet high and coming in tassel, the lateral spread of roots has a radius of about 4 feet and penetrates the soil to a depth of 3 to 4 feet. The number of roots continues to increase until the plant is mature, when they fully occupy the upper 3 to 4 feet of soil.

The depth to which roots may penetrate is somewhat dependent on the character of the soil, as is shown by the Colorado station. In a black adobe soil, the roots were limited mostly to the upper 12 inches, while on another heavy soil containing much clay they penetrated only 24 inches.

13. Distance from surface. — At a distance of 6 inches from the plant the upper roots are usually about 3 inches below the surface, sloping gently to 4 or 5 inches deep at a distance of 2 feet from the plant. However, when there is abundance of moisture in the surface, feeders may come within 2 inches or less. Distance from the surface seems to be controlled by the presence of sufficient moisture, and also by the degree of shading, since roots are very sensitive to light. Late in the season, when the soil is well shaded, roots will be found very near the surface; but ordinarily, during the growing season, they are 3 to 4 inches below. The method of planting may also exercise some influence on the depth of upper roots. At the Kansas station,[1] where the root systems of "listed" corn were compared with those of surface-planted, the upper roots of the former were found to average about 1 inch deeper during the cultivating season, especially near the plant, thus permitting deeper cultivation.

14. Types of roots. — Maize roots may be classed as primary roots, brace roots, lateral roots, and hair roots. The main roots are those having their origin at the base of the stem; they are twenty to thirty in number and 4 to 6 feet in length. The lateral roots are numerous small roots thrown off from these, and they again may produce other laterals. Their number is very large and may average several hundred to each main root; in length they vary from less than 1 inch to 1 or 2 feet. The root-hairs are microscopic in size, single-celled, and infinite in number. They are borne on the main roots in their earlier growth, and on all the laterals. Root-hairs are short-lived and limited to the newer root growth, or rather to a zone near the growing point of the roots. They are absorbent organs, and do not grow to be roots.

15. The proportion of root. — The total weight of the root in a corn plant has been found to be about 12 to 15 per cent of the weight of the total plant, including the ear.[2] The total length of roots laid end to end, of a single plant of small grain, as wheat or oats, has been estimated at 1600 feet; but in a corn plant it would be greater.

[1] Kan. Agr. Exp. Sta., Bul. *127*: 203.

[2] Kiesselbach. Nebr. Agr. Exp. Sta., Rpt. 1910: 131.

16. The amount of root. — The amount of root developed is more or less in response to the needs of the plant. When moisture is abundant or excessive, the plant will not develop so much root as when the moisture content is normal or below normal. Also in very dry soil, with a moisture content below the wilting point of plants (about 12 per cent in loam soils), the growth of roots is limited, as is also the case when the soil is very hard.

17. Functions of the root. — The root functions may be stated as: (1) the absorption of water and of salts in solution; (2) the excretion of organic substances, especially carbon dioxid, and possibly free organic acid, also mineral salts and the salts of organic acids; (3) the solvent effect of the excretions on soil particles.

The absorption of water and solutions, as well as the exudations, take place largely through the root-hairs. These root-hairs are constantly produced from the epidermal cells near the growing root tip. They are forced into close contact with the soil grains; in fact, the soil grains are more or less embedded in the root-hair tissues. Each soil grain in a moist soil is surrounded by a film of water containing more or less mineral matter dissolved from the soil. This soil water is absorbed by the root-hair, and it seems probable that exudations from the root-hair also aid in freeing less soluble minerals in the soil grains. The process of absorption is by means of osmosis.[1]

[1] *Osmosis.* — When two solutions of different density are separated by a porous membrane, there will be first a movement of the weaker solution through the membrane into the stronger, and later a return movement, the process continuing until the two solutions have the same density. The contents of a root-hair being denser than the soil solution surrounding it, there is a constant movement of the soil solution into the root-hair. By some means the *exosmosis*, which would take place in the case of an ordinary membrane (movement of the cell solution outward), seems to be restrained in the root-hair, probably by some functional

18. The stem. — The stem of maize differs from that of other cereals in the fact that it is solid — filled with pith — while others are hollow. The maize stem may vary in height from 2 feet, in the case of dwarf pop corn, to 18 or 20 feet in some of the tall southern varieties.

The nodes not only serve to strengthen the stem, but are also the points of origin for all its lateral outgrowths, as roots, branches (tillers), leaves, and ears.

The stem usually extends not more than three to five inches below the ground surface. This part is divided into about six to ten short nodes, each bearing a whorl of roots. Above the soil surface each node bears a leaf and in addition either a branch or an embryonic ear. The early northern varieties of maize, with a height of about 6 feet, usually have about eight to ten nodes above the soil, while the tall southern varieties may have eighteen to twenty. A typical plant in Illinois or Indiana will have about fourteen nodes, with one or two branches from the surface nodes and an embryonic ear at each node; usually, however, only the ear at about the eighth node develops, the others remaining dormant.

In Fig. 13 is shown a stem from a plant about 10 inches high. The full number of nodes, and also of leaves, is formed. Growth of the stem from this point on will be by a lengthening of the internodes, but there will be no increase in number of nodes. This is called internodal growth, in distinction from the apical, or terminal, growth of many other plants — as peas and beans, where new growth is constantly taking place at the apex.

The outer part of the stem is a thin shell of hard tissue,

activity of the cell. The result is a much greater movement into the root-hair than exudation out of it. The soil solution passes from the root-hair into the root and is finally transmitted to the stem and leaves.

the function of which is to give strength and rigidity. A cross-section of the stem will show, in addition to the pith,

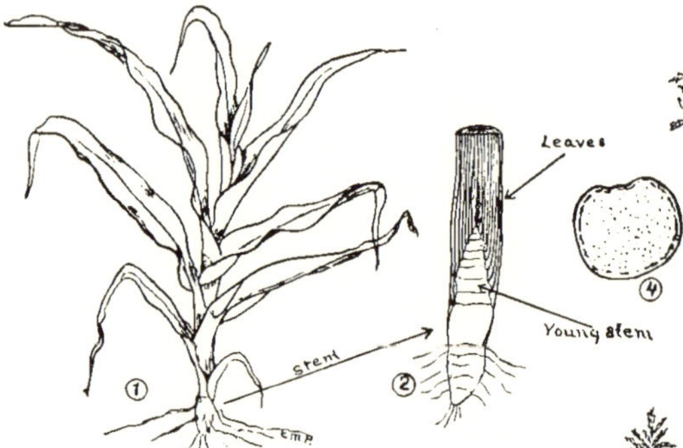

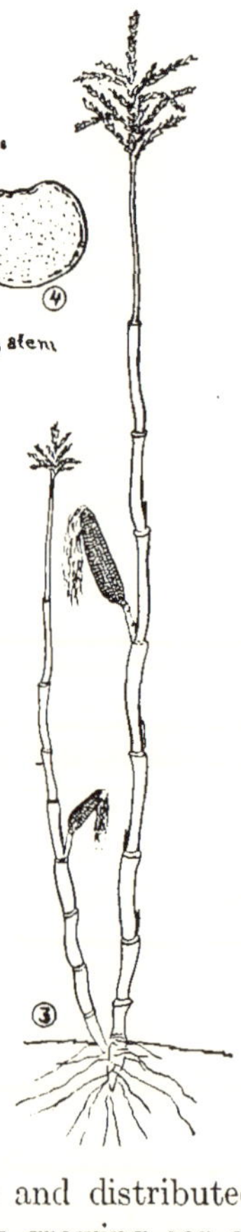

FIG. 13. — Development of the corn stem. 1. Plant about 10 inches high. 2. Section of 1, at base, showing that all nodes, leaves, and tassel are more or less developed at this stage; growth is internodal. 3. Full-grown stem with leaves removed. 4. Cross-section of stem.

a large number of fibrous strands, known technically as *fibro-vascular bundles.* It is through these bundles that the water taken in by the roots passes up the stem and is distributed throughout the plant; and again, when the leaves have elaborated plant-food from the material taken up from the soil and out of the air, this plant-food is carried down these same fibro-vascular bundles and distributed to those parts where it is needed, as the growing ear or the roots.

19. Tillers. — If a young corn plant about 8 inches high is carefully dissected, two or more small buds will be noted in the axils of the first leaves. If conditions are favorable, one or more of these buds will develop into a branch of the plant, or a "tiller." If conditions are unfavorable, as in poor soil, or when the plants are close, the buds may remain suppressed and never grow. On a cold clay or wet soil very few of the tillers develop; while on a warm, sandy soil, especially if fertile, every plant may develop one to three or four tillers. A good example of this is the very abundant tillering common in cornfields in the light but fertile soils on the west edge of the corn-belt (central Nebraska); while the same varieties on the heavier clay soils of Ohio or New York will rarely develop tillers. Every corn plant has several latent buds, which may develop if conditions are favorable, but which otherwise may remain dormant. The tiller may develop its own root system and ears, and may function in all respects as a normal plant. A tendency to tiller, however, is somewhat hereditary, as certain small varieties of flint and sweet corn normally produce well-developed ear-bearing tillers, while some of the large dent varieties seldom tiller.

20. Leaves. — If a small corn plant a few days old be taken and a cross-section made just above the first node, the full number of leaves may be identified, wrapped into a kind of stem (Fig. 13). As the stem elongates the leaves are gradually exposed, but the leaf growth takes place mostly while the leaves are yet enfolded. There is very little increase in size after the leaf is fully exposed.

The *structure* of a leaf is more complicated than appears from a casual examination, because of its many functions. The functions are principally: (1) to provide for the free circulation of solutions and air throughout the leaf; (2) to give off constantly large quantities of excessive water taken up by the roots; (3) to elaborate plant-food from the minerals and water taken out of the soil, combined with carbon and oxygen taken from the air; (4) to absorb energy from the sun which is necessary in order that

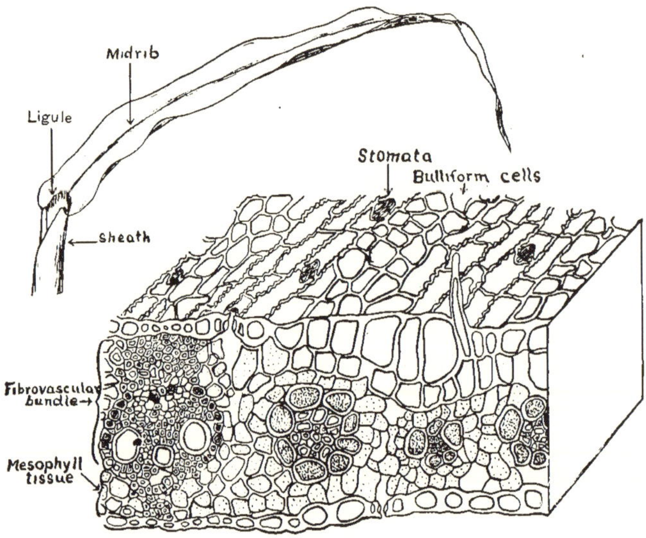

FIG. 14. — Leaf structure. The movement of water and solutions takes place through the fibro-vascular bundles. The mesophyll tissue furnishes the means for elaborating plant-food from raw material. Interchange of air and gases takes place through the stomata. The bulliform cells are similar to mesophyll cells, but contain a large percentage of water. Shrinkage of these cells causes the leaf to roll in dry weather.

these activities may proceed. Each of the above functions of the leaf requires specialized tissues which are briefly described as follows (21, 22): —

21. The vascular system. — If a maize leaf is examined, there will be found running lengthwise a large number of parallel veins. On examining a cross-section of the leaf under the microscope, each vein will be seen to contain a fibrous bundle of various kinds of tissues, known as a *fibro-vascular* bundle. In Fig. 14 are shown some of these large, thick-walled cells, resembling somewhat the veins of an animal; and it is by means of these that solutions are circulated through the leaf. These fibrous bundles extend into the stem and the roots, making a direct passage for the transfer of soil solutions taken up by the roots through the stem and out into the leaves.

22. Air passages. — Throughout the leaf tissues are systems of air passages. These are connected with small openings of the leaf surface, or *stomata*. Fresh air is constantly coming into the leaf through these stomata, carrying carbon dioxid and oxygen, both of which are utilized by the plant in connection with the minerals taken up from the soil and elaborated into plant-food.

23. Loss of water. — As the air passes out of a leaf it constantly carries out the water that has been taken up from the earth. The outer covering, or epidermis, of the leaf is impervious to water or air, but there are stomata at regular intervals. The number of these is very great,

Kind of Leaf	Number of Stomata in One Square Inch	
	Upper Side	Under Side
Indian corn (*Zea Mays*)	60,630	101,910
Sunflower (*Helianthus annuus*)	112,875	209,625
Red clover (*Trifolium pratense*)	113,515	216,075
Hop (*Humulus hupulus*)	0	165,120
Apple (*Pyrus Malus*)	0	156,670
Pea (*Pisum sativum*)	65,145	139,320

usually being most numerous on the under side. The above table gives the number estimated for several kinds of leaves.[1]

The stomata also close more or less when the leaves begin to wilt, thus preventing to some extent the loss of moisture.

24. Chlorophyll-bearing cells. — The work done by the leaf involves the expenditure of energy. There are a large number of cells in the maize leaf filled with minute green bodies, called chlorophyll grains. These not only give the green color, but arrest the energy of the sun's rays, making use of this energy to perform the various activities of the plant.

25. The flower. — The male, or *staminate,* flowers are borne in the tassel. The anthers are three in number and filled with pollen. While the pollen sacs are small, about one-fourth inch in length, yet each is estimated to contain 2500 pollen grains.

The female, or *pistillate,* flowers are borne on the ear and are closely related in structure to the male flowers. When very young, they are borne in pairs, but one is very small and seldom develops. Occasionally both of these grains develop in the tassel flowers of pod corns. Sturtevant[2] mentions also an ear of podded flint corn from Ohio, in which the kernels were twinned. These reversions indicate that at some time in the early evolution of maize both these flowers functioned, but for some reason only one now develops.

The principal parts of the pistillate flower are an ovary, or egg cell, a carpel which surrounds this for protection, and a long extension of the carpel, called the style, or "silk." The details of fertilization are given later.

[1] Bessey, C. E. Botany (Briefer Course), p. 45.

[2] Bul. Torrey Bot. Club, *1894*: 336.

26. The ear. — The probable origin of corn from some grass-like plant similar to teosinte is discussed under Biological Origin (p. 15).

The ear may be regarded as a branch of the main stem, the ear stem having exactly as many nodes as the main stem above the ear, the husks corresponding to the leaf sheaths and the ear to the tassel; the side branches, however, are no longer present, while the central spike has been enlarged into a cob, and the pistillate flowers, or grains on the ear, correspond to the pollen flowers.[1]

The ear is the storehouse of the maize plant, where is produced not only the young germ, but also a store of starch, protein, oil, and other products for its future nourishment, much as a swarm of bees makes a store of honey for the young, laying eggs in the cells at the same time. As mentioned heretofore, these products are first prepared by the leaves and later transmitted to the ear.

References on distribution of maize roots: —

PAMMEL, L. H. Grasses of Iowa. Iowa Agr. Exp. Sta., Bul. *54* : 8–13.

KING, F. H. Wis. Agr. Exp. Sta., Rpt. *1892* : 112; and *1893*: 160.

HAYS, W. M. (1889.) Minn. Agr. Exp. Sta., Bul. 5.

TEN EYCK, A. M. (1899–90.) N. Dak. Agr. Exp. Sta., Bul. *36*: 43.

SHEPPERD, J. H. (1905.) N. Dak. Agr. Exp. Sta., Bul. 64.

TEN EYCK, A. M. (1904.) Kan. Agr. Exp. Sta., Bul. 127.

Colo. Agr. Exp. Sta., Rpt. *1896* : 181.

N. Y. Agr. Exp. Sta., Rpt. *1888* : 171.

References on tillering of maize: —

Neb. Agr. Exp. Sta., Bul. *91* : 16.

Pop. Sci. Mo., Jan. 1906 : 55.

[1] What is an Ear of Corn? Pop. Sci. Mo., Jan. 1906.

CHAPTER IV

PHYSIOLOGY OF CORN PLANT

PLANT physiology deals with the activities and functions of the physical parts of the plant. Not all parts of a plant have a present important function. Certain parts may be regarded as rudiments left in the process of evolutionary change, and they may even be detrimental. In other cases, certain parts may be regarded as only chance variations of no value from an economic view point. It is therefore important to make a careful analysis of plants, to determine the function of each part, which parts have an important function, and how the proper activities of the plant are favored or hindered.

27. Living plants. — One of the distinctive characters of living plants as compared with dead material is the fact that many forces of nature may act as a " stimulus " and get a response entirely at variance with the usual result. This is well stated in Strasburger[1] as follows: —

"The free end of a horizontally extended flexible rod bends downwards merely by its own weight. The same result will follow if any part of a dead plant, such as a dry stem, be substituted for the rod. But if a living, growing stem be used in the experiment, then the action of gravity will manifest itself in a manner altogether at variance with its ordinary operation. That part of the stem which is still in a state of growth will ultimately curve upwards, *and by its own activity assume an up-*

[1] STRASBURGER, NOLL, SCHENK, and KARSTEN. (1908.) Textbook of Botany, p. 173.

right position; it moves in a direction exactly opposite to the attractive force of gravity. If a tap-root be similarly experimented upon, it will, on the contrary, continue its downward movement until it places itself in a line with the direction of the attraction; a rhizome, however, under like circumstances, would constantly maintain its growing apex in a horizontal position. In these three experiments, the force of gravity is exerted upon horizontal portions of plants. The physical conditions are the same in each case, yet how entirely different the results."

The above phenomena are some of the manifestations of "life." In the same way, light, heat, moisture, and other physical factors will act as a "stimulus" to living plants, but the response is not always what would be obtained with dead material, and it may be the opposite. This fact should be kept in mind in dealing with living plants.

28. Stability of the plant. — A corn plant one inch in diameter at the base may be 100 to 125 inches in height, yet it will have a broad spread of leaf, bear a heavy ear, and be able to maintain itself without breaking or falling prostrate in a heavy wind. A rye plant bearing a heavy head may be five hundred times as tall as the diameter of its base. This rigidity of the plant body is necessary in order that it may reach considerable height and expand its leaves to light and air. Rigidity is due principally to *turgidity* in the soft tissue or young plant, and to the *mechanical tissues* in the older and stronger parts.

29. Turgidity. — In the leaves of a corn plant is a certain set of cells, known as bulliform cells. These are located near the upper surface between the ribs, or veins. (See Fig. 14.) When moisture is abundant, these cells absorb water until they are turgid. The leaf is then spread out flat and is more or less rigid and brittle. However, when the weather is very hot or when soil moisture is low, the cells lose water enough so they are no longer turgid, and the leaf then becomes limp and rolls up. In the same way, all cells of the plant may be more or less turgid, aiding in giving rigidity to the plant body.

30. Tension. — If a section a few inches long of the stem of green corn be taken and the outer peripheral tissue be removed from the pith, the pith will at once expand in length and some force will be required to restore it to normal length. It will thus be seen that there is a natural tension at all times between the outer cortex and the pith. This tension adds to the rigidity of the stem.

31. Mechanical tissue. — The supporting framework is made up of woody and fibrous tissues in the outer part and the nodes of the stem and in the midribs and veins of the leaves. These are mostly comprised of fibers (sclerenchyma or bast) of great tensile strength. Quoting from Strasburger, "the sustaining strength of sclerenchymatous fibers is, within the limits of their elasticity, in general equal to the best wrought iron, or hammered steel." The fibers are bound together, giving a strong elastic body.

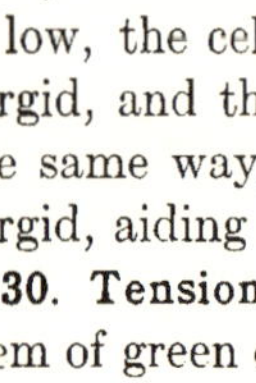

Fig. 15. — Illustrating resistance to bending when the supporting tissue is on the outside of the stem, as in corn. One side must be shortened and the other stretched.

The location of the framework on the outside, rather than in the center, of the stem adds to the rigidity. For example, if an elastic rod be bent (Fig. 15), the inner side is shortened and the outer lengthened. If a supporting skeleton be placed in the center of this rod, then the rod is flexible and considerable bending would be possible without much resistance from the center; but if the sup-

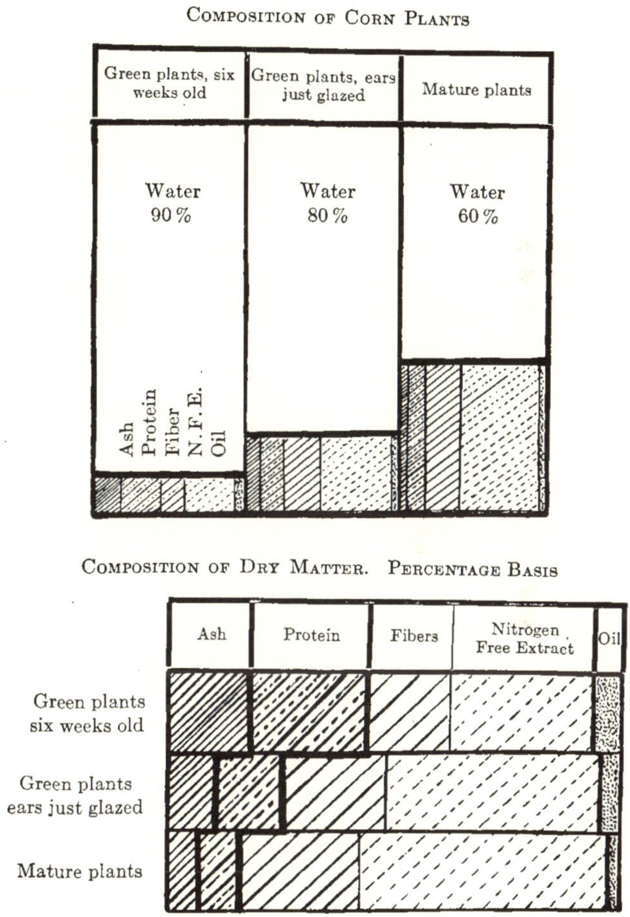

Fig. 16. — Composition of corn plants at three stages of growth. The upper figure shows a progressive increase in percentage of dry matter as the plant approaches maturity. The lower figure shows the progressive change in character of dry matter.

porting skeleton be on the outside, then much greater resistance is offered.

In the root, however, the mechanical tissue is in the center, thus allowing the root to bend easily about among obstructions and at the same time giving pulling strength.

32. **Nutrition.** — Probably the most interesting, as well as the most important, knowledge regarding the corn plant is the method by which its supply of elements for growth is secured from the soil and air, and the factors affecting the assimilation and use of such plant-food.

THE COMPOSITION OF A CORN PLANT

33. If a green corn plant 2 or 3 feet in height be dried in an oven until all the water has been driven out, it will be found that about 90 per cent of the total weight is water and only 10 per cent is solid, or dry matter. When in the roasting-ear stage, the plants are about 80 per cent water, and later, at maturity, 60 per cent.

TABLE IX

Average Composition of Green Maize[1]

Constituent	Composition of Fresh Plants	Composition of Dry Matter at Three Stages		
		Six Weeks Per Cent	Ears just Glazed	Mature Plant
Water	79.0			
Ash	1.2	17	10	6
Protein	1.7	27	14	9
Fiber	5.6	17	22	25
Nitrogen free extract	12.0	35	50	57
Fat	.5	15	4	3

[1] From Jenkins and Winton, Office of Exp. Sta., U. S. Dept. Agr., Bul. 11, 1892.

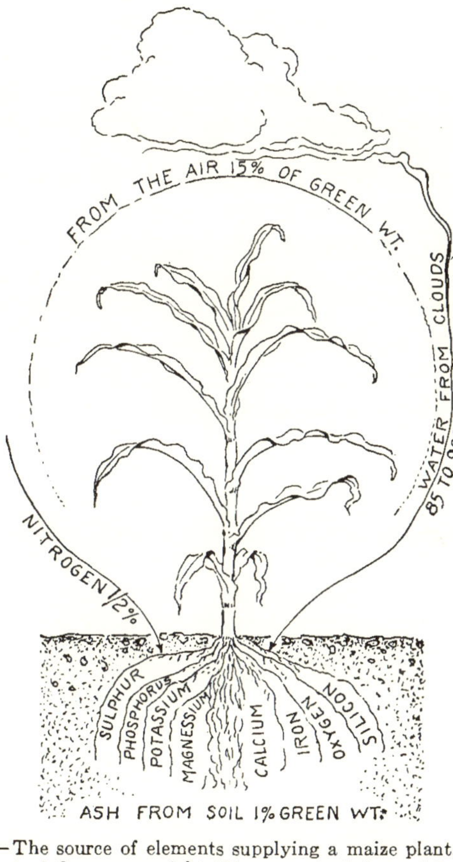

FIG. 17.—The source of elements supplying a maize plant. About 10 per cent of the green weight (50 per cent of dry weight) is carbon. About 5 per cent is oxygen and 4 per cent hydrogen. The oxygen is derived from the air in combination with carbon, from water or from oxide salts. The hydrogen comes principally from water. Ash from the soil equals about 1 per cent, and nitrogen ½ per cent of the green weight. About 80 per cent of the green weight is water, not in composition.

The dry substance is combustible, and when it is ignited, about 90 to 95 per cent will be consumed, leaving a residue of ash. The combustible part consists principally of the elements *carbon, hydrogen,* and *oxygen,* with a smaller quantity of nitrogen. The ash left is made up of mineral substances taken from the soil. Thus, only about one per cent of the weight of a green plant comes from the soil.

34. The essential constituents. — There are ten essential elements necessary to plants, one of these coming from the air, two from water, and six from the soil, while one — nitrogen — comes indirectly from the air through the soil. *Carbon* comes only from the carbonic acid gas of the atmosphere, *hydrogen* and *oxygen* from water (oxygen also from the air, and oxid salts), *nitrogen* from the soils, as nitrates or ammonium salts. The other six essentials, namely, *sulfur, phosphorus, potassium, calcium, magnesium,* and *iron,* are taken from the soil.

Plants do not find these elments in simple forms, but in combination — for example, the hydrogen and oxygen from water, where it is in combination as H_2O, and carbon from carbon dioxid (CO_2). All the minerals, as phosphate, potassium, and the others, are always found in combination. A demonstration of how plants can live on these minerals when in solution may be made by taking pure distilled water and dissolving the following mineral salts (after V. D. Crone) :—

Distilled water	1–2 liters
Potassium nitrate	1.0 gram
Ferrous phosphate	0.5 gram
Calcium sulfate	0.25 gram
Magnesium sulfate	0.25 gram

If properly handled, a corn plant may be grown to maturity in this solution.

In addition to the "essential" elements found in the ash of plants there are also other elements, as sodium and silicon, found in large quantities; but these are probably not essential to growth.

THE ABSORPTION OF WATER

35. It has been pointed out in the text that the water absorbed by plants is a dilute solution of all the soluble substances in the soil, the absorption taking place through the vast number of root-hairs, from which the water solution passes into the lateral roots, up the stem, and out into the leaves. The water passes up the fibrous bundles found all through the pith. This can be demonstrated by cutting off a stem near the ground early in the morning, when root-pressure is high. Water will soon exude in small drops wherever the fibro-vascular bundles are cut. During the heat of the day, root-pressure is negative, and no result can be secured.

THE GIVING OFF OF WATER

36. Water loss[1] from the plant serves several functions, the most important of which is the concentration of the water solution. By constant evaporation of water the salts taken up in solution are left in the plant, to be utilized in its growth. The leaf is so constructed as to facilitate the giving off of quantities of water and at the same time protect the inner tissues.

The leaf is covered with a strong epidermis, which has, however, an enormous number of stomata. The number in a single corn leaf of average size is estimated at sixteen to twenty millions. These small openings are connected

[1] Water loss from the plant is of two kinds, namely, *transpiration* and *evaporation*. The former is closely associated with assimilation, and the amount of water given off as a result of this process is comparatively small. The greatest loss is by simple evaporation, in common with all objects exposed to dry air.

with a series of air spaces in the leaf so that there is free movement of air into and out from the leaf. Also, the vascular bundles, which deliver the water from the roots into the leaf, are spread out in the leaf into a fine network so that every part is quickly supplied with water as it evaporates.

The quantity of water evaporated from day to day depends directly on the conditions of climate and on the amount of leaf area exposed. An average corn plant has about *8 square feet* of leaf surface, while a full stand of corn has a total leaf area equal to twice the area of land on which the corn is growing; in other words an acre of land would have about *two acres* of leaf surface. The daily water loss per plant varies from *3 to 10 pounds*, depending on the humidity of the air and on the wind, just as does any other object or a free water surface. The following data, taken at the Nebraska Experiment Station, illustrate the above statements: —

TABLE X

DAILY VARIATION IN WATER LOSS FROM PLANTS AND FREE WATER[1]

Interval of 24 Hours, ending at 7 P.M.	Water Loss per Plant Grams	Water Loss from Free Water Grams
July 27, 1911	4550	454
July 28, 1911	2333	372
July 29, 1911	1579	173
July 30, 1911	2802	232
July 31, 1911	3561	314
August 1, 1911	3982	374
August 2, 1911	3419	311
August 3, 1911	2143	204

[1] Nebr. Agr. Exp. Sta., 24th Ann. Rpt., p. 102. 1911.

ASSIMILATION

37. The taking up of carbon from air and uniting it with other elements to form plant tissues is called assimilation. Carbon is found in nature, as coal or graphite, or it is artificially prepared from wood, as charcoal. Carbon is the most important constituent of all plants, composing about 50 per cent of the dry weight.

Carbon can be demonstrated by charring, that is, by burning a piece of maize stem without sufficient air for complete combustion, when other substances will be driven off by the heat, leaving the carbon. So much carbon is present that the stem will retain its shape and structure.

When any substance is burned or decomposes, the carbon present passes into the air as carbon dioxid (CO_2). This gas constitutes about 0.03 per cent of the atmosphere.

A maize plant takes air into the leaves through the air pores (stomata) and extracts the carbon dioxid. The air then passes out again, carrying water and by-products — often oxygen — of which the plant should rid itself.

38. The necessary energy for maintaining the activities of the leaf is derived from the sunlight. Some of the leaf cells contain small green *chlorophyll* bodies. When the plant is in strong sunlight, these chlorophyll bodies rapidly accumulate starch grains. If the plant is placed in darkness, however, no starch will be made.

In the same way, we may show the necessity of carbon dioxid, by placing growing plants in air artificially freed of this gas. Even in the presence of bright sunshine, no starch will be accumulated.

39. The by-product of assimilation is pure *oxygen.* The chemical process of the manufacture of starch from carbon dioxide and water, through the activities of chloroplasts, may be illustrated as follows, leaving out intermediate steps —

$$6\ CO_2 + 6\ H_2O = C_6H_{12}O_6 + 6\ O_2$$
(glucose + oxygen)

$$C_6H_{12}O_6 = C_6H_{10}O_5 + H_2O$$
(starch + water)

While the starch is made in the leaves it cannot be distributed in this form to other parts of the plant, as starch is insoluble. It is therefore first converted into sugar and in this form is distributed to the stem, roots, ear, or wherever needed for growth. The juice of a green maize stem may contain 10 per cent or more of sugar during the earing season, when it is being transported from leaves to stem and ear. This soluable sugar may be converted into many forms of carbohydrate material, as fiber or starch. In the ear it is principally deposited again as starch.

40. Growth. — Following the plan outlined by Sachs, the growth of a maize plant may be divided into three distinct phases, as: —

1. The early growth period (embryonic), in which the rudiments of new organs are formed.
2. Elongation of the already formed embryonal organs.
3. Period of internal development.

The first period covers about the first three weeks of growth. A plant three weeks old will have all parts, as the full number of leaves and nodes, most of the main roots, and embryonic tassel, ears, and tillers. From this time on, growth consists principally of the elongation and development of these parts. Later there is a third phase, that of internal development, as the depositing of starch in the ear and the strengthening of fibrous tissues.

REPRODUCTION

41. In maize, as in most plants, nature has provided for the perpetuation of the race through the abundant production of seeds. An average maize plant produces

FIG. 18. — An ear of corn in full silk, just ready for pollination.

about 1000 seeds, usually all on one ear; in some varieties, however, two or more ears are produced.

42. Pollen. — The pollen, or fertilizing element, is produced in the tassels and usually begins falling one to two days before silking; there is great irregularity in this

FIG. 19. — The process of fertilization of the corn flower. Each embryonic grain produces a long style or "silk." Each silk must receive one or more pollen grains.

respect, however, some plants producing silk before pollen.

43. Style. — Each grain produces a style, or silk. The grains about one-fourth of the way from the base of the ear silk first, and the process passes gradually toward the tip, the entire period of silking requiring two to four days.

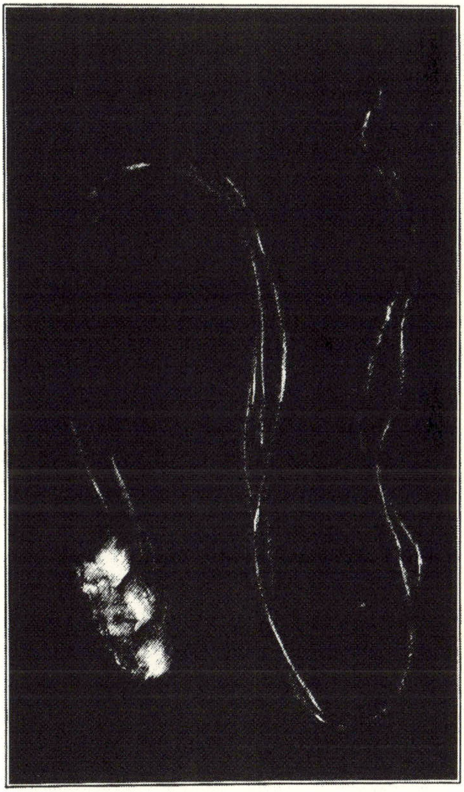

FIG. 20. — Young corn kernels and silks.

44. Fertilization. — For fertilization to take place, every silk must receive at least one pollen-grain, and fertilization is probably surer if it receives several. As the pollen is distributed by wind, it must be very abundant to insure pollination ; therefore, ten to twenty thousand pollen grains are produced to every ovary, or embryonic kernel.

The exposed end of the silk, or style, is covered with fine hairs and is also adhesive, so that pollen-grains readily adhere when they come in contact. It is not necessary for the pollen to fall on a particular part of the silk; it may reach any of the exposed surface. In fact, fertilization has been accomplished by placing the pollen on the silk within the husk.

Soon after a pollen-grain falls on a receptive silk it sends out a tube, or filament,

FIG. 21. — Ear of corn showing zone poorly fertilized. The ear silks in successive zones from near the butt toward the tip. Some unfavorable condition happened when this zone was in silk.

which penetrates the silk; and soon the contents of the pollen-grain pass down to the egg, in the embryonic seed at the base of the silk. Immediately upon fertilization, the ovule begins a rapid growth. In case a part of the silk should fail to receive pollen, those ovaries will not develop, and the result will be irregular rows on the ear. Sometimes in very hot and dry weather the pollen is killed and will not fertilize. Also, insects such as grasshoppers often eat off the silks, or a part of them, thus preventing fertilization.

Several investigators have studied fertilization and embryonic development of the corn ovule, as Guignard, Webber, True, and Poindexter. No one has reported observing the passage of a pollen-tube down the silk. There is some question as to whether the pollen-tube actually passes down within the tissues of the style, or whether it may not follow the slight depression or groove on one side of the style. Guignard calls the opening near the base of the style the "stylar canal," and thinks that the pollen-tube enters this opening, but he did not observe it. When the ovule is finally reached, it has not been definitely observed at just what point the pollen-tube enters.

TRUE, RODNEY. Bot. Gaz. *18*: 215.

POINDEXTER, C. C. The Development of the Spikelet and Grain of Corn. The Ohio Nat., Vol. IV, No. 1, Nov. 1903.

SECTION II

PRODUCTION AS RELATED TO CLIMATE AND SOILS

CHAPTER V

RELATION OF CLIMATIC FACTORS TO GROWTH

The ability of corn to yield is indicated by certain maximum yields that have been obtained under favorable conditions. Edward Enfield,[1] in 1866, listed a number of record yields which had been published between 1840 and 1866.

45. The average of fourteen record yields collected from seven States was 145 bushels per acre, two of these records being 200 bushels per acre. The American Agriculturist[2] records in 1857 a yield of 857½ bushels on 5 acres, or an average yield of 171½ bushels per acre. Hartley[3] reports a 90-acre field of corn in Pennsylvania averaging 130 bushels per acre, the same farmer having averaged 100 bushels per acre for twelve years. The four largest yields on record are as follows: —

Year	Grower	Place	Yield Bushels
1857 .	Dr. J. W. Parker	Asylum Farm, Columbia, S.C.	200.3 (*a*)
1889 .	Capt. Z. J. Drake	Marlboro, S.C.	255.1 (*b*)
1889 .	Alfred Rose	Yates Co., N.Y.	213.0 (*c*)
1910 .	Jerry Moore	Winona, S.C.	228.7 (*d*)

(*a*) This record has often been mentioned, but original data to verify it are not available. (*b*) and (*c*) These records and the method of growing are given in The American Agriculturist, *XLIX*, March, 1890, p. 122. In each case the yield is field weight at husking and would have to be reduced at least 10 per cent for crib dry weight. (*d*) Field weight.

[1] Enfield, Edward. (1866.) Indian Corn, p. 54.

[2] The American Agriculturist, *XVI* : 238. 1857.

[3] Hartley, C. P. (1910.) U. S. Dept. Agr., Farmers' Bul. *414* : 14.

In all of the above cases, enormous quantities of commercial fertilizers and manures were used, but the instances illustrate the ability of corn to yield under the most favorable soil conditions.

The possible yield of corn if all conditions, both climatic and soil, were ideal for a season, is probably in advance of any yield thus far recorded. The average yield of corn in the United States is 26 bushels per acre, only a small proportion of the possible production.

CLIMATIC FACTORS AND GROWTH

46. The principal elements of climate are sunshine, heat, rainfall, humidity, and wind.

The climate favorable to corn is determined not so much by the amount as by the distribution of these factors, without fluctuations so great as to retard the growth or to

Fig. 22. — A single corn plant bearing 5 ears. Demonstrating the productivity of corn in favorable environment.

reduce vitality.[1] For example, one section might have sufficient average rainfall for a normal crop, but if this rainfall so fluctuated that at one season it was excessive and at another deficient, the normal crop might be reduced one-half or more; while another region with no more total rainfall but a better distribution would have a normal crop. In the same way, a single frost out of season or a hot wind might do great damage, although the average temperature might appear favorable. Average annual rainfall, temperature, and sunshine are not a safe guide, unless the fluctuation of these factors during the growing season is also known.

47. Length of the growing season. — Corn differs somewhat from other cereals in being able to adjust itself to the growing season. Wheat, oats, and barley grown in northern regions yield as well as when moved farther south, or even better. They have a somewhat longer growing season when taken south, but do not occupy the available period as does corn. Some North Dakota varieties of corn will mature in 80 days, while Gulf States varieties often take 200 days. There are large corn regions with a growing season of more than 200 days, but it does not appear that corn has been able in any region to utilize to advantage a longer growing period.

As the tropics are approached, while frosts cease to limit the crop-growing season, at the same time there is usually a dry period which serves as a limit. In Mexico the growing season is limited in this way.

All other factors being favorable, we may assume that

[1] The effect of fluctuation of rainfall on crop production is discussed in Bul. 130, Bur. Plant Indus., U. S. Dept. Agr., 41–49, in an article on "Cost of Crop Production under Humid and Dry Conditions."

the ability of corn to yield will increase with the length of the growing season up to somewhere near 200 days. Therefore, for a good share of the present corn-belt of the United States, the length of the growing season is an important limiting factor. However, the varieties most

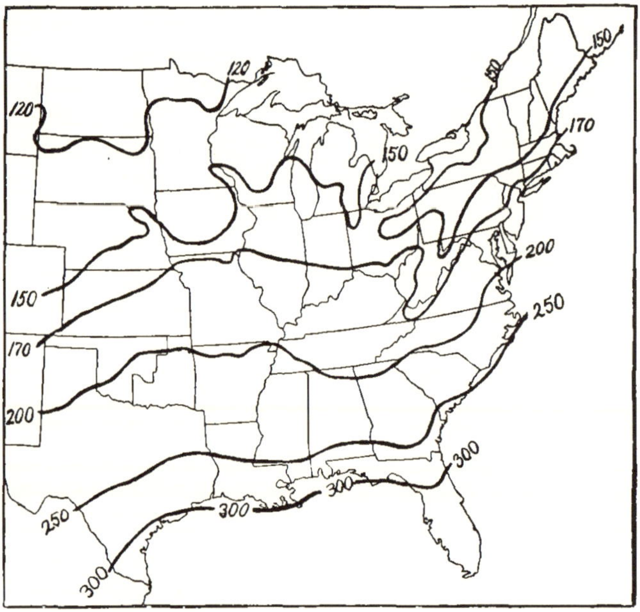

Fig. 23. — Length of growing season as indicated by the average date of last killing frost in spring and first killing frost of fall. (Bul. V., U.S. Weather Bureau.)

commonly grown in the South mature in 160 to 180 days, due to other limiting factors than frost, such as the rainfall not being sufficient for the entire season, poor drainage in early spring, or an unfertile soil.

The accompanying chart, taken from Bulletin V of the

United States Weather Bureau, shows the average length of the crop-growing season, or rather the time between the average date of the last killing frost in spring and the average date of the first killing frost in fall.

The growing season of corn nearly coincides with the last probable frost of spring and the first probable frost of fall. For example, at Lincoln, Nebraska, where the average time between killing frosts is given as 165 days, it is not considered advisable to grow a variety of corn taking more than 130 days to mature. The growing season for corn would be, in general, 20 to 30 days less than indicated on the chart; or the 200-day limit would be central South Carolina.

Also, there is great fluctuation in the length of growing season from year to year at any one point, and there is a general tendency to grow corn that will mature in the shortest season. Frear[1] made a study of meteorological conditions in relation to the development of corn at the Pennsylvania station for three ears, 1887 to 1889. In his conclusions he makes this statement: "The difference in temperature between these two seasons (1887 and 1889) is almost equal to the difference in the mean July temperature of Quebec and Boston; of Burlington, Vermont, and Philadelphia; and of Fort Assiniboine, on the northern boundary of the United States, and Santa Fé, New Mexico. Then, too, in 1889 the rainfall was almost twice as great as in 1887, and the cloudiness at least 25 per cent greater."

48. Relation of sunshine to growth. — The function of sunlight in furnishing the necessary energy for the various activities of plant growth was discussed in the

[1] FREAR, W., and CALDWELL, W. H. Relation of Meteorological Conditions to the Development of Corn. Penn. Agr. Exp. Sta., Ann. Rpt. *1889*.

chapter on physiology. This has been well expressed by Abbe, as follows:[1] —

"The growth of a plant and the ripening of the fruit is accomplished by a series of molecular changes in which the atmosphere, the water, and the soil, but especially the sun, play important parts. In this process a vital principle is figuratively said to exist within the seed or plant and to guide the action of the soil, the water, and the air into such new chemical combinations as will build up the leaf, the woody fiber, the starch, the pollen, the flower, the fruit, and seed. . . . No plant life, not even the lowest vegetable organism, is perfected, except through the influence of the radiation from the sun. . . . The radiation from any artificial light, especially the most powerful electric light, will accomplish results similar to sunlight; therefore it is not necessary to think that life, or the vital principle, is peculiar to or emanates from the sun, but on the contrary that living cells utilize the radiations or molecular vibrations so far as possible to build up the plant."

49. The intensity of sunlight. — The intensity of the sunlight received on the earth's surface is modified by the altitude of the sun, which determines the total hours of sunshine duration, by the atmosphere, and by the clouds.

At high noon on a perfectly clear day, if there were no atmosphere, the earth's surface would receive the full effect of the sun's rays. When the sun is at zenith the atmosphere absorbs about 12.5 per cent of the sun's energy, so the efficiency may be expressed as .875, assuming the full effect to be 1. However, the sun is only at zenith for a moment, therefore, as it approaches the hori-

[1] ABBE, CLEVELAND. Relations between Climates and Crops. U. S. Weather Bureau, *1905* : 15.

zon, the altitude decreases, until at the horizon the light must penetrate 12 to 35 times as much atmosphere, and its total effect is weakened to about one-fifth the full effect at 90 degrees. The effect at different altitudes is expressed in the following selected altitudes:[1] —

TABLE XI

Altitude of Sun Degrees	Thickness of Atmosphere Laplace Formula	Intensity of Direct Sunshine; Calories per Minute, per Sq. Cm. on Surface Normal to Rays
0	35.50	0.359
4	12.20	1.293
10	5.70	1.868
30	1.995	2.275
50	1.305	2.364
90	1.000	2.403

From the equator to 40 degrees latitude, the total sunshine received at a given place from March to September is about one-third of the total possible sunshine at that point if the sun stood at zenith during the hours of daylight. In the northern latitudes the longer days of midsummer compensate for the lower altitudes of the sun, so that during the months of June and July, as much heat is received at the north pole as at any lower altitude. In fact, for a period of about 90 days, more heat units are received at the north pole than the equator, but due to the great amount of ice is not sufficient to raise the temperature above freezing. The relative quantities of heat received at different latitudes in the Northern Hemisphere are shown by the following table, as calculated by Aymonnet:[2] —

[1] Abbe, *loc. cit.*, p. 85. [2] *Ibid.*, p. 92.

TABLE XII

Month	Latitude						
	0°	10°	30°	50°	70°	80°	90°
March 20–31 . .	3.7	3.7	3.3	2.3	1.1	0.6	0.2
April	10.0	10.6	10.1	8.0	5.4	3.9	3.4
May	9.8	10.7	11.7	10.5	9.0	8.6	8.7
June	9.2	10.4	11.9	11.3	10.7	11.0	11.1
July	9.7	10.7	12.1	11.3	10.3	10.1	10.2
August	10.1	10.7	10.9	9.2	6.8	5.9	5.8
September 1–23 .	7.7	7.8	7.1	5.2	2.7	1.5	0.9
Total . . .	60.2	64.6	67.1	57.8	46.0	41.6	40.3
Total possible if sun stood at zenith . .	186.0	186.0	186.0	186.0	186.0	186.0	186.0

It is apparent from the above data that up to 70 degrees north latitude there is sufficient sunshine during the summer months to produce corn, were it not for other limiting factors, as low temperature due to a cold soil and cold air currents.

The data presented thus far are on the basis of perfectly clear days, but the presence of clouds reduces the sunshine. At Montsoris, France, careful records for the corn-growing season kept from 1875 to 1885 showed only about 40 per cent of the possible intensity of sunshine, due to cloudiness. Corn under such conditions does not grow well, but requires, even at that latitude, what might be termed a rather " sunny " climate.

We may conclude that except where cloudiness prevails for half the time, there is sufficient sunshine for corn production even up to 70 degrees latitude.

50. Relation of rainfall to growth. — The transpiration of 14 to 20 tons of water is required to produce one bushel of corn. For a yield of 50 bushels per acre, this equals 7 to 10 acre-inches of water.[1] With a larger crop the water used would be increased proportionally. Under field conditions there must be added to this whatever loss may take place through run-off, evaporation from the soil, and seepage. King found that a yield of 7000 to 8000 pounds

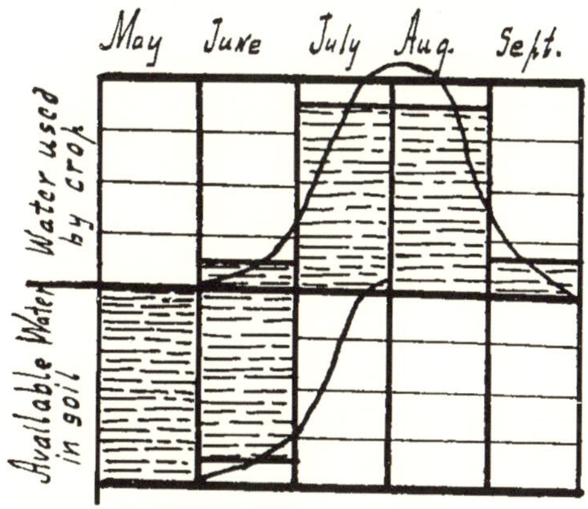

Fig. 24. — Chart showing relation between storage water in the soil and consumption of water by the corn plant each month. The storage capacity of the soil is exhausted before the end of July. The crop is therefore dependent on July and August rainfall.

of dry matter per acre (approximately a 50-bushel yield) required about 12 acre-inches under field conditions. In this case the loss by seepage, run-off, and evaporation must have been about 5 acre-inches (assuming 7 inches used by the crop), but this will vary with the soil, cultivation, distribution of rainfall during the growing season, and amount of storage water in the soil at planting time.

[1] Montgomery, E. G. Ann. Rpt. Nebr. Agr. Exp. Sta. *1901* : 155. King, F. H. Ann. Rpt. Wis. Agr. Exp. Sta. *1902* : 99.

An average corn soil in good tilth will store about 5 to 6 inches of available water in the upper 4 feet. A 50-bushel crop would then require at least 6 inches additional rainfall during the growing season, and probably more than this, as corn seldom grows well when required to exhaust the soil moisture to low limits. A 75-bushel crop would require an additional rainfall of 10 inches and a 100-bushel crop at least 15 inches during the growing season, in addition to that stored in the soil. When the run-off is large, as on hills or with torrential rains, or when there is seepage, the above estimate should be increased. This estimate is on the assumption that the soil is fertile. No amount of rain would make a poor soil productive. For example, the average rainfall for June, July, and August in the eight surplus corn States is about 12 inches, but the average yield is 28.5 bushels. Other factors than total rainfall here limit the yield, one important factor being that the rainfall is not always properly distributed.

51. Any system of culture that will serve to prevent run-off on the one hand and to decrease evaporation on the other, will proportionally increase the available water supply for the crop.

Not only the total amount, but the distribution, of the season's rainfall is of great importance. Figure 25 shows the precipitation for June, July, and August for a period of fifteen years and the yield for eight surplus corn States, namely, Ohio, Indiana, Illinois, Iowa, Nebraska, Kansas, Missouri, and Kentucky.[1]

Here is shown a very close relationship between rainfall and yield, when large areas are considered.

[1] SMITH, J. WARREN. Relation of Precipitation to Yield of Corn. U. S. Dept. Agr. Year Book, *1903*: 215-224.

Professor Hunt,[1] at the Illinois Agricultural Experiment Station, grew 18 plats of corn which yielded 32 bushels per acre. The next year, and on the same plats and with the same varieties of corn, the yield was 94 bushels per acre. The rainfall from May to September was 13 inches the first season and 22.5 inches the second season.

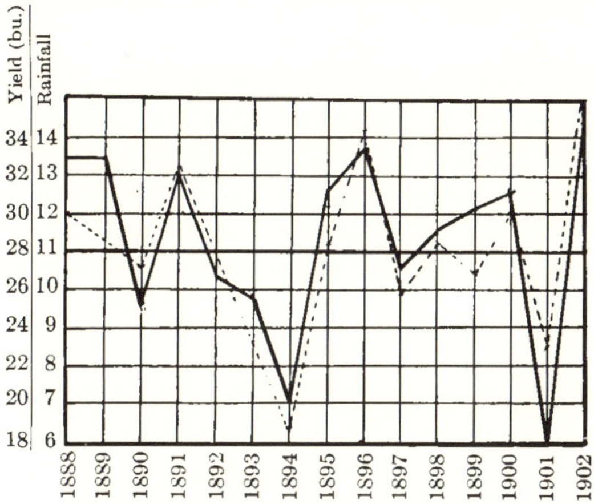

FIG. 25. — Rainfall of June, July, and August, and yield of corn per acre. (Year Book, U.S. Dept. Agr., 1903.)

——— Average yields of corn 1888 to 1902.
------ Average rainfall for June, July, and August.

The seasonal rainfall and its distribution is the most important climatic factor in corn production. With sufficient rainfall, *properly distributed,* it is probable that the present yield of corn would be increased 50 to 100 per cent. We cannot control the rainfall or its distribution during the season, therefore farm practice must make the best use of rainfall as it comes. The present rainfall is sufficient for two to three times the present yield, if it is conserved and the soil is in the most fertile condition.

[1] HUNT, T. F. Cereals in America, p. 207.

CHAPTER VI

RELATION OF SOILS TO GROWTH

MOST of the good corn soils of the United States are deep black loams, well drained, well supplied with organic matter, and rich in available nitrogen, phosphates, and potassium.

52. The soil may be regarded as a medium for holding minerals and water in an available form for the plants as needed. Natural productive soils are those that in a state of nature contain all the mineral elements and organic matter necessary, and are supplied with sufficient natural rainfall.

In some virgin soils, as the deep black loam soils of the Mississippi, Ohio, and Missouri river drainage basins, there is sufficient of all mineral elements in an available form for the maximum production of corn. Even in these soils, however, maximum production is seldom attained, as the rainfall is not always properly distributed, nor even sufficient.

Corn especially enjoys a large supply of nitrogen and will flourish in soils so rich in available nitrogen that other cereal crops would produce an excessive amount of straw, probably lodging and making a poor yield of grain. Corn is able to make use of fertility furnished through the decaying of coarse organic matter, as manure or sod land; while other cereals, as wheat and oats, require for best results a more advanced state of decomposition, with the elements more easily available.

The ability of corn to utilize to advantage large quantities of fertilizer and manure is illustrated in the cases cited on page 57 of the four maximum yields of corn produced.

Fig. 26. — Corn as it grows on the best type of natural corn land.

CAUSES OF LOW PRODUCTION

53. Assuming rainfall to be sufficient, a good corn soil should produce 75 bushels per acre. Only a small percentage of the corn land in the United States will yield this at present, due to certain causes which may be summarized as follows: —

1. Poor drainage. Corn suffers more than do other cereals from poor drainage, as it requires a "warm" soil, and also available nitrogen in rather large quantities. Nitrifying processes are hindered in waterlogged soils.

2. Surface soil depleted through erosion, very common on rolling lands in regions of large rainfall.

3. Soil once fertile but depleted through constant cropping without return of organic matter or minerals.

4. Soils which in a virgin state were deficient in organic matter or lacking in some mineral element.

Each of the above soils will be found deficient in one or more of the following: —

(*a*) Drainage.

(*b*) Organic matter.

(*c*) Nitrogen.

(*d*) One or more mineral elements.

(*a*) is corrected by drainage, (*b*) and (*c*) by manure or the growing of legumes, (*d*) by manure or commercial fertilizers.

CLASSIFICATION OF CORN SOILS IN THE UNITED STATES ACCORDING TO PRODUCTIVENESS

54. For the regions east of the Rocky Mountains the corn soils may be classed according to productivity into four general groups.

1. Soils capable of producing 75 bushels or more per

acre, with normal rainfall of region such as the black loam bottom land soils of the Mississippi drainage basin, and certain areas of black upland or drained swamps. This soil is well drained, well supplied with organic matter, minerals, and rainfall, and usually commercial fertilizers will show little or no effect. The total area is small, probably not greater than 1 per cent of the Corn Belt. This may be termed the ideal corn soil.

2. Soils producing 35 to 50 bushels per acre, with favorable climatic conditions.

(*a*) This includes the greater part of the cultivated lands in the surplus corn States of Ohio, Indiana, Illinois, Iowa, Nebraska, Kansas, and Missouri. These soils have been cropped for fifty to seventy-five years, during which time the ability to yield has decreased 25 to 50 per cent. All these soils respond quickly to an application of manure, or are increased 25 to 50 per cent in productivity by growing a crop of clover or alfalfa. They seem to need organic matter and available nitrogen more than anything else. The supply of minerals is generally sufficient, but in many cases the application of both potassium and phosphates gives increased yields, though, as a general rule, the increase is not sufficient to be profitable. Rotation, the use of legumes, and manure are to be relied on at present as the principal means of maintaining or increasing the yield.

(*b*) All the "good corn land" through the Eastern and Southern States is also included in this class.

3. Land producing 25 to 35 bushels under favorable climatic conditions.

(*a*) Through the Eastern and Southern States are large areas which are fairly productive when first brought under cultivation, but which have been cropped for seventy-five years or more. Erosion also has played an important part in depleting the rolling lands. The supply of organic matter is generally low, and in many cases the lands need underdrainage. Throughout the "Corn Belt" there are also considerable areas in this class.

(*b*) Soils naturally not very productive, through lack of one or more mineral elements or of drainage.

In general, legumes and manure must be the principal means of increasing and maintaining the productivity of this land; but when a mineral element is lacking, as lime, potassium, or phosphorus, it will usually be necessary to add this in the form of commercial fertilizer.

4. Land producing less than 20 bushels per acre.

(*a*) Through the Eastern and Southern States are large areas which, through continuous cropping and erosion, are low in yield. In addition to the prevention of erosion, the same general treatment as is recommended for the previous class may be used.

(*b*) Land in regions of deficient rainfall. Where there is less than eight inches during the growing season, lack of moisture becomes a limiting factor in corn production. From Dakota to Texas there is a large area with a fertile soil but an annual rainfall of only 18 to 25 inches. In these soils conservation of moisture is the most important phase of soil treatment.

SUMMARY

55. The ability of corn to yield is indicated by certain maximum yields, when 150 to 200 bushels per acre have been harvested. Regarding climatic factors, there is usually enough sunshine and, in most of the Corn Belt, a sufficient total rainfall; but the latter is not often distributed in the best way for the growth of corn. A large share is lost by run-off, and the supply is seldom properly conserved by preparation of the land and by cultivation.

The length of growing season is a limiting factor, where the season is less than 180 days.

The principal cause of low production is lack of available fertility in the soil.

Climatic factors are mostly out of our control except that the effect of rainfall may be modified, hence our principal efforts in increasing corn production should be in the treatment of soil.

SECTION III

IMPROVEMENT AND ADAPTATION OF THE CORN PLANT, AND ENVIRONMENT

CHAPTER VII

EARLY CULTURE OF CORN

INDIAN corn was unknown to Europeans until the discovery of America. At that time it was found to be in general cultivation by the Indians of both North and South America. In fact, corn was the principal crop cultivated by the native Americans, as they had neither oats, wheat, nor barley, and very few of the cultivated vegetables. The most ancient evidence of the culture of corn is found on the western coast of South America and in Mexico. In Peru specimens of corn have been found in connection with ancient ruins or geological formations, which are probably at least two or three thousand years old. The fact that corn was buried in the tombs, as well as other evidence, indicates that it had an important place in the religious ceremonies of this semicivilized people and was probably their most important cultivated plant.

56. During the fifteenth century the earliest white explorers of America took corn back to Europe, where in time it came to be extensively cultivated, especially in those countries surrounding the Mediterranean Sea.

When corn culture began to spread in Europe it had many curious names, as Italian corn, Turkish corn, Spanish wheat, Guinea wheat, and others, probably indicating the places where its culture first became extensive.

Collins has recently described a type of corn cultivated

in China. References to corn in Chinese literature indicate its culture in China for some 350 years, although just how or when corn was introduced into China is a question.

When the first white settlers came to America, at Jamestown (1607) and Plymouth (1620), they at once took up the culture of corn, procuring the seed and learning the method of culture from Indians. It soon became the most important cereal crop of the colonists, gaining its popularity by reason of its simple culture, its sure production, and the ease with which the crop was harvested and preserved.

DEVELOPMENT OF VARIETIES

57. In 1898 Sturtevant listed 507 named varieties and 163 synonyms. It was not possible for Sturtevant to secure all varieties in his day, and it is probable that a complete catalogue of all varieties at present would almost double this number. Of these varieties listed by Sturtevant, 323 were classified as dent corn, 69 as flint corn, 63 as sweet corn, 27 as soft corn, and 25 as pop corn.

It is known that at least a few varieties of all the five principal groups were in cultivation when America was discovered, with the possible exception of sweet corn. The earliest record we have of sweet corn is in 1779, when it was mentioned as being in cultivation near Plymouth, Mass.[1] However, it could easily have been overlooked by the early explorers and has probably been in existence for a long period.

It appears that the Indians inhabiting what is now the northern part of the United States and southern Canada

[1] STURTEVANT, E. L. U. S. Dept. Agr., Office of Exp. Sta., Bul. *57* : 18.

cultivated mostly an eight-rowed flint corn, and in a limited way an early variety of soft corn commonly known to-day as "squaw" corn. The Indians of the southwestern United States, Mexico, and South America cultivated the different varieties of soft corn principally, and also, in a limited way, flint corn, pop corn, and dent corn. The dent corn, however, does not appear to be like our modern dent of the deep-grained, large-eared varieties, such as Boone County White, but of a rather shallow-grained type with a square grain or a grain even broader than long. There was also a very rough, deep-grained type with a short ear, similar to our Shoe Peg corn of the present day.

By the year 1800 there were a number of recognized varieties of flint corn, mostly of eight-rowed types, and a few dent and soft corns cultivated by the colonists. At least one variety of sweet corn (the Papoon eight-rowed) and a few pop corns were known, but were not in general cultivation.

Bonafous, in 1836, and Metzger, in 1841, both published classifications and descriptions of corn indicating that at least all the characters of corn known at present were to be found among the varieties at that time. Metzger made twelve races, and mentioned varieties ranging in height from 18 inches to 18 feet. Since 1840 there has been a rapid expansion of corn culture and great interest has been shown in the development of varieties adapted to various conditions and uses. It may be safely estimated that perhaps three-fourths of the present varieties of corn have been developed since 1840. The history of sweet corn is an excellent example. Following are listed the authorities and the number of varieties of sweet corn that each knew, and the year of his observation: —

Date	Authority	Number of Varieties
1779	. . . .	1
1832	Bridgman	1
1836	Bonafous	1
1853	U. S. Patent Office Rpt.	3
1858	Klippert	6
1866	Burr	12
1884	Sturtevant	33
1898	Sturtevant	63

There has been a similar rapid development of dent varieties.

In 1866 Edward Enfield[1] made a list and description of corn varieties, which he said represented " most of the varieties in use, and all that are likely to be of practical value to the farmer." Of field corns he describes 20 varieties, 13 of which were flints, 3 broad-grained dents of 12 or 14 rows, 1 flour corn, and 3 gourd seed varieties grown in the South. However, J. S. Leaming had begun selecting " Leaming " corn in 1826 and Mr. Reid began selecting his corn in the late forties. With the rapid development of corn culture, after the Civil War, the modern dent type came to be generally used throughout the Corn Belt States; although the flint corns are still the principal corns in the Northern States, where the season is too short for the large dents.

58. Early methods of modifying varieties. — Probably the means that has been most commonly used in the past was either to hybridize two varieties and select some type from this hybrid for several years until the type was fixed,

[1] Enfield, Edward. (1866.) Indian Corn. D. Appleton & Co., New York.

or to start a systematic selection in some recognized variety. One of the earliest reports of the origin of a variety is given by Mr. C. H. Heydrick, of Utica, Penn., in the Agricultural Report of the Commissioner of Patents for 1853. As most early varieties were originated by some such method, the quotation is here given: —

> "With regard to the changes which may be wrought in a variety by cultivation, I cannot give a better illustration than the history of the 'Vermont Yellow,' that I cultivated a few years ago. Its characteristics were, a short stalk, slender above the ear, strong below, ears small, with eight rows, thick at the butt end, growing near the ground, and frequently having a stem two feet in length. My plan of selecting seed from this variety was to choose from such stalks as produced two or more ears, rejecting those with large butt ends, and such as were not set close to the stalk. Such seed was hard to find the first year. The second year nearly one-half of the stalks produced two ears, and there were fewer long stems and large butt ends. A milder climate had also produced another change. Many ears appeared with ten or twelve rows. This induced me to improve the size of the corn and accordingly I selected as before, adding such ears as contained more than eight rows, together with a few ears of a larger sort. Continuing this system for a few years I obtained a variety characterized by the following marks: stalks light, seldom exceeding six feet in height; strong below the ears, slender above; ears containing from ten to fourteen rows, and from two to three ears to a stalk, more frequently than a less number. From these facts it will be seen that a mixed variety may be produced, possessing all the desirable qualities of several old ones. But such a new variety will require attention a few years, to prevent it from degenerating into one of the original sorts, after which, I think, the variety will become as permanent as any other."

Many of the early corn breeders used the above method of selecting out some type from an old variety. It is probable, however, that many of the variations found were due to natural hybridization, as this is likely to take place to some extent in a neighborhood where any two fields are less than twenty to forty rods distant from each other.

FIG. 27. — Relation of type to climate. The short type on left is better adapted to dry regions than the tall, more slender type. The tall type is adapted to warm, more humid regions.

Crossing as a means of securing new forms was often practiced, and a method is outlined by Enfield (1866).[1]

[1] ENFIELD, EDWARD. (1866.) Indian Corn, pp. 70–74.

59. Natural selection and acclimatization in producing varieties. — It is well known that each region of the United States has corn of a type more or less peculiar to that section. For example, in the Gulf States, corn grows very tall, frequently 15 to 18 feet, with ears 6 or 8 feet from the ground; in the Corn Belt States the plant is about two-thirds as high; while along the Canadian border the height is 5 to 8 feet, and ears are often less than 2 feet from the ground. Also the growing season will vary from 200 days in the South to 80 days in the North. If a variety of corn be moved from one section to another, it will become from year to year more like the native corn of the region.

It is not known how much of this change may be due to actual modification of the plant by environment, but it is probable that it is brought about chiefly through wide variations and through both natural and artificial selection. When a variety is moved from one climate or soil to another, it does not yield so well the first year as later, when it becomes "acclimatized." When planted first in the new location, there are certain plants much better suited than others to the new conditions. These would produce the best ears and be selected for seed, thus preserving the best-adapted type. An excellent example is cited from Nebraska,[1] where a variety of corn from Iowa was grown in central Nebraska for two years and as a result decreased about 12 inches in height, while the ear was almost 8 inches lower; the yield of grain, however, increased.

If the same variety be widely distributed and grown for a few years, and seed again collected for comparison under the same conditions, it will be found that

[1] Nebr. Agr. Expr. Sta., Bul. *91* : 29.

each region has had some effect in modifying the original type.

SUMMARY

60. The early culture of corn probably originated in the high plateau region of southern Mexico, about the beginning of the Christian Era. From here it spread north and south, its culture being general throughout North and South America by the year 1000 A.D.

The Indians grew flint and flour corns chiefly, probably because of their keeping and germinating qualities.

Most of the modern varieties in general use in North America have been developed during the past century, though the principal types have probably been in existence for many hundred years.

Selection, both artificial and natural, accounts for the origin of many varieties; while crossing, sometimes intentional, but often accidental, has furnished a great many variations from which to choose.

References on early culture: —
See, References on early history; p. 24.

References on origin of varieties: —
Mo. Agr. Exp. Sta., Bul. *87* : 113.
Nebr. Agr. Exp. Sta., Bul. *83* : 12.
U. S. Dept. Agr. Yearbook, *1907* : 331.
BOWMAN AND CROSSLY. Corn, p. 424.

CHAPTER VIII

IMPROVEMENT OF VARIETIES

PERHAPS no other cultivated plant in America has been the object of so much study and attention, with the object of adapting it to the various soils, climates, and needs of man, as the corn plant.

The plant is large, interesting, lends itself well to detailed study, and responds readily to care or selection. From earliest domestication by white men, there has always been a large number of growers, giving time and attention to its improvement. An infinite variety of forms has been developed. Every detail of the plant has been studied as to its possible economic value, in improving the yield or quality of grain or forage. Almost every possible theory has been held by practical growers regarding the relative value of different types of ear, leaf, stem, or other parts of the plant. Of recent years, good scientific study has also been made at many of the experiment stations. In the following pages it is attempted to sum up what is known to be of practical value in type of plant or selection of methods.

61. Type of ear. — From the earliest times it is probable that some attention has been given to the types of ear selected for seed. The originators of varieties have usually had a well-defined type in mind, for which they have selected. There is no evidence that the type of ear chosen has had a direct relation to yield, since equally good results have been secured with very diverse types, as the tapering Leaming, the cylindrical Reid's Yellow Dent, the shallow Hickory King, and the extremely deep-grained Hackberry. Flint corns and the small-eared, prolific corns have also given excellent results.

Since several investigators have studied ear characters in relation to yield, all data verify the experience of Hartley, which he summarizes as follows: "A careful tabulation of yields as compared with other ear characters, covering six years' work with four varieties, embracing in all more than 1000 ear-to-row tests of production, indicates that no visible characters of apparently good seed ears are indicative of high yielding power." Since white men began corn culture, no doubt some gain has been made in ability to yield, but the fact that large ears have been selected will account for this. As the ear represents the producing ability of a plant, all other things being equal, the selection of large ears would preserve the most productive strains.

Varieties having a medium depth of grain mature better and keep better in the crib than very deep-grained types, and, since they seem to yield as well, they are to be preferred.

62. Type of plant. — While some study has been given to the character of plant, no definite relationship has been proved, which would justify the consideration of the plant in seed selection, other than this: *the average type in an acclimated variety will yield better than either extreme.* The average type, however, varies in different regions. For example, on the west edge of the Corn Belt, with a rainfall of 22 to 25 inches (central Nebraska, Kansas, and vicinity), the plant when acclimated is short, stocky, with the ear rather low. To select here for tall plants with the ear borne high would not be in harmony with natural conditions, and experience has shown that varieties having these characters do not yield so well as the native type. Also, at the Nebraska station, when comparison was made between broad-leaved and narrow-leaved strains in dry years, there being an excess of sunshine and limited water supply, the largest yields were obtained from narrow-leaved strains.[1] In one year, with excessive rainfall, broad-leaved types gave larger yields. While broad leaves elaborate starch, they also evaporate water at a rapid rate; hence, the most desirable leaf area on corn plants must be a balance between the moisture supply on the one hand and sunshine on the other.

While under rather abnormal conditions for growth, as a dry climate, attention must be given to the character of plant growth, yet under normal conditions a wide range is permitted. In Illinois, where selection for height of ear was continued for six years, there was no marked difference in yield between the high-ear and low-ear types, and the same was also true when angle of ear was considered.

It may be assumed that when selection is made for yield, all other characters of the plant will adjust themselves under the given conditions; so that ultimately the type of plant giving best yield under those conditions will result. However, under certain conditions in the South the ears are borne very high, and in the North, in some cases, the ears are borne very low. In both these cases, for convenience in harvesting, it would be well to select for a more desirable height of ear. There are many instances where selection to modify some character of the plant would be justified, even though the yield was not affected.

[1] Nebr. Agr. Exp. Sta., 24th Ann. Rpt., p. 158.

SYSTEMS OF SELECTION

63. Mass Selection in corn is the method of selecting from a large field a number of individuals that conform nearest to some ideal type. Seed of these plants is mixed together and planted a second year and again a large number of ears are selected and mixed for planting another year, and so on for many years.

It was discovered, however, that of two ears much alike in appearance, perhaps one might yield 25 per cent to 50 per cent more than the other when used as seed corn. The importance of testing each ear separately was at once recognized.

In *pedigree selection* after the first mother ears are selected, a separate record is kept on the performance of each ear or its progeny. For example, if it is desired to select for a type bearing ears low on the stalk, a hundred such ears might be selected from a large field. If mass selection is practiced, they are mixed together and planted the following year and the method continued for several years. If pedigree selection is followed, each ear is planted in a separate row and a record made of the percentage of low ears produced by each mother ear. Seed ears are only saved from those mother ears producing a large percentage of low ears, the remainder being discarded. Perhaps ten mother ears out of the first 100 will be found to transmit the desired quality. From the progeny of these ten ears, 100 ears may again be saved, each to be planted in a separate row. This may be continued for several years, the performance record being kept for every year. By keeping a record on each family separate it will be possible to gradually discard those families not transmitting the desired quality and keep only those that are most desirable.

64. Results with mass and pedigree selection. — With all obvious characters, such as height or angle of ear, the same results to a certain degree will be obtained by either method; but these results should be secured in less time by the pedigree method.

This is rendered more comprehensive by conceiving a cornfield to be a mixture of types, and selection as a

FIG. 28. — Selection for high and low ears at the Ill. Exp. Sta. The tape shows height of ear. The original selection was from the same field and had been continued for five years when this picture was taken.

method of isolating these types.[1] It is clear that pedigree breeding is a more rapid method of isolation than continuous selection.

With selection for yield it is possible that no result would be secured with mass selection, unless there were some obvious character of the plant closely associated with yield; while on the other hand, rapid progress might be expected from pedigree cultures, as a record would be kept on the performance of each individual.

[1] The theory of isolating pure types by mass and pedigree selection is expounded at length by De Vries, in Plant Breeding. Open Court, Chicago.

65. Mass Selection. — The result of mass selection in corn is well illustrated by the history of any of the older varieties, as Leaming, Reid, or Boone County White. After many years' selection, the breeder succeeded in producing a more or less uniform type.

For example, the Leaming variety was originated by Mr. J. S. Leaming, of Hamilton, Ohio:[1] "After fifty-six years' selection, Mr. Leaming produced a corn having as variety characteristics a distinctly tapering ear, fairly large butts, rather pointed but well-covered tips, with kernels of a deep yellow color, with very irregular rows."

Hartley produced a corn with twisted rows by selecting such ears from the field. At the Nebraska Agricultural Experiment Station, a shallow-kerneled type of corn was fixed by continuous selection after five years.[2]

66. Pedigree selection. — A striking example has been reported from the Illinois station.[3] Two sets of Leaming

TABLE XIII

General Averages of Crops produced in Corn Breeding, for High Ears and for Low Ears

Year	Height of Ear		Height of Plant	
	High-ear Plat Inches	Low-ear Plat Inches	High-ear Plat Inches	Low-ear Plat Inches
1903 . . .	56.4	42.8	113.9	102.5
1904 . . .	50.3	38.3	106.2	97.4
1905 . . .	63.3	41.6	128.4	106.5
1906 . . .	56.6	25.5	116.3	86.0
1907 . . .	72.4	33.2	130.4	99.7
1908 . . .	57.3	23.1	114.0	79.3

[1] Mo. Agr. Exp. Sta., Bul. *87* : 113.

[2] Nebr. Agr. Exp. Sta., Bul. *112* : 20.

[3] Ill. Agr. Exp. Sta., Bul. 132. 1909.

ears, one borne high on the stalk and the other low, were selected in the fall of 1902. Continuous selection was practiced for six years, when a difference of about three feet in height was secured in the average crop, as shown by the table on previous page.

Results were also obtained when selection was made to increase or decrease the angle of the ear, the erect-ear strain and declining-ear strain having average angles of 46° and 88.50° respectively, after six years.

67. Selection for composition. — Composition can also be modified by continuous selection, as shown by the Illinois station.[1] Ears of a single variety were selected for high-protein, low-protein, high-oil, and low-oil content, respectively. After ten years' selection, the high-protein

TABLE XIV

TEN GENERATIONS OF BREEDING CORN FOR INCREASE AND DECREASE OF PROTEIN AND OIL

YEAR	HIGH-PROTEIN CROP PER CENT	LOW-PROTEIN CROP PER CENT	DIF-FER-ENCE	HIGH-OIL CROP PER CENT	LOW-OIL CROP PER CENT	DIF-FER-ENCE
1896	10.92	10.92		4.70	4.70	
1897	11.10	10.55	0.55	4.73	4.06	0.67
1898	11.05	10.55	0.50	5.15	3.99	1.16
1899	11.46	9.86	1.60	5.64	3.82	1.82
1900	12.32	9.34	2.98	6.12	3.57	2.55
1901	14.12	10.04	4.08	6.09	3.43	2.66
1902	12.34	8.22	4.12	6.41	3.02	3.39
1903	13.04	8.62	4.42	6.50	2.97	3.53
1904	15.03	9.27	5.76	6.97	2.89	4.08
1905	14.72	8.57	6.15	7.29	2.58	4.71
1906	14.26	8.64	5.62	7.37	2.66	4.71

[1] Ill. Agr. Exp. Sta., Bul. 128. 1908.
U. S. Dept. Agr., Farmers' Bul. *366* : 314.

selection contained almost twice as much protein (14.26 to 8.64) as the low, while the high-oil selection contained almost three times as much oil as the low (7.37 to 2.66).

At the Nebraska station, pedigree selection for yield was practiced for five years in one case and two years in another, and an increased yield, amounting to nine bushels was secured in both cases.

TABLE XV

(CLASS I) RESULT OF FIVE YEARS' PEDIGREE SELECTION FOR YIELD. BUSHELS PER ACRE

	1907 Bushels	1908 Bushels	Average Bushels
Selected strains	82.0	66.0	74.0
Check plats (original stock)	72.5	59.0	65.7
Difference	9.5	7.0	8.3

(CLASS II) RESULT OF TWO YEARS' PEDIGREE SELECTION FOR YIELD

	1908 Bushels
Selected strains	68.0
Check plats (original stock)	59.0
Difference	9.0

References on type of ear and stalk: —

Ohio Agr. Exp. Sta., Bul. 212. (1909.)

Nebr. Agr. Exp. Sta., Bul. No. 91, p. 12 (1906); No. 112, p. 17 (1909).

Nebr. Agr. Exp. Sta., Ann. Rpt. 1910, p. 154.

Cornell Bul. 287. (1910.)

Ill. Bul. 132. (1909.)

HARTLEY, C. P. Yearbook U. S. Dept. Agr. 1909, pp. 309–320.

References on methods of continuous pedigree selection: —
Ill. Exp. Sta., Buls. 74, 82, 100, 128.
Conn. Exp. Sta., Buls. 152 and 168.
Ohio Exp. Sta., Circ. 66.
Nebr. Exp. Sta., Bul. 112.
Directions to Coöperative Corn Breeders. Hartley, C. P. (1910.) Bur. of Plant Industry, Wash., D.C.

CHAPTER IX

METHODS OF LAYING OUT A BREEDING-PLAT

While the principle underlying systematic selection is simple, it has been more difficult to develop good methods for carrying out the selection work in order to avoid error. Breeding-plot methods have had a steady development, each step in advance being intended to overcome some source of error or to develop some new possibilities.

In the reference on "Methods of pedigree selection," given on a previous page, several methods for conducting a breeding-plat are given. The development of the breeding-plat plan may be summarized in the following brief way, beginning about 1895: —

68. 1. **Select a number** of mother ears and plant in parallel rows, taking the yield of each row and saving seed ears from this row to continue in the same manner. (See Ill. Agr. Exp. Sta., Bul. 55.)

2. The above plan was found to favor inbreeding and close breeding, with danger of decreasing yield. In order to avoid this it was recommended to detassel every odd row, sowing seed from only the detasseled rows. Thus, every odd row became a dam while the even rows would be the sires. By duplicating the plat and detasseling the even rows in the duplicate, seed could be saved from every mother ear. (Ill. Agr. Exp. Sta., Bulls. 82 and 100.)

3. About 1904 to 1906, Professor Williams, of Ohio, applied the "check row" system to the breeding-plat and developed the "ear remnant" plan. The "check row" was a composite planted for every sixth row. This was for the purpose of checking the uniformity of the land, as the breeder might unconsciously select for a breeding-plat a piece of land more fertile at one side than at the other.

One difficulty found with the ear-to-row method in practice was that the best-yielding row might chance to be between two very poor yielders, so that seed ears saved from this row would be partly crossed with the poor-yielding rows on each side. In order to meet this difficulty, the plan was adopted of planting only a part of each ear, sufficient to determine the kind of progeny it would develop, and the remainder of the ear was kept. The next season, remnants of only the best ears would be planted in parallel rows and a part detasseled; but with the poor-yielding rows eliminated, all the fertilization would come from desirable rows. (See Ohio Agr. Exp. Sta., Circ. 66 (1907); Amer. Breeder's Assoc., *III:* 110.)

HOW TO CONDUCT A BREEDING-PLAT

69. As there is considerable inquiry at present regarding methods of corn breeding, it seems best at this time to outline a plan which experience so far seems to recommend.

Variety to use. — Select some variety that is well adapted to the region and is a good yielder. This is important, as one might spend years in working on a poor variety, and in the end have nothing better than the best variety already existing. It may be well to do some preliminary variety testing.

Selecting the ears. — If yield is to be the principal object of selection, it will not be necessary to hold closely to some one type of ear. In fact, since we do not know definitely what particular type of ear in a variety may do best in a new locality, it would seem wise to select several types, the main consideration being that the ears are sound and well matured.

Number of ears to select as foundation stock. — Exceptional ears are not common, there being probably not more than one in every fifty to one hundred ears. Therefore, if one starts with only a small number of ears, twenty-five to fifty, he may not find a single exceptional yielder in the lot. Not less than one hundred ears, and preferably two hundred should be tried out in the preliminary trial.

The test plat. — Great care should be exercised in procuring a uniform piece of land for the test plat, as everything depends on being able to compare in an accurate way the yields of the different ears. The land should not be exceptionally rich, but only of the average fertility of the region. If the land can be plowed twice — say fall-plowed, and then backset in the spring — and disked several times, this will do much toward equalizing conditions.

Size of plat. — Half an ear will plant a row 16 to 20 rods in length. However, there will be less error if the rows are duplicated, and it is best to plant two rows 8 rods long from each ear. One hundred ears will make two hundred plats 8 rods long. This will take a piece of land 32 by 11 rods or 16 by 22 rods; or two test plats one-half this size on different parts of the farm may be used, duplicating the experiment in each.

Check plats. — No matter how carefully the land is selected, it may lack uniformity; for this reason, check plats should be planted with a uniform lot of corn. It has been found very satisfactory in practice to make every fifth plat a check. The simplest way is to make a check of every plat that is a multiple of 5, as 5, 10, 15, and so on.

Planting the ears. — The land should first be laid off by a marker into checks 3 feet 8 inches apart. The planting must be done by hand. Carry the ear, and shell off the grains as needed. It is best to plant four grains in a hill; then, when the corn is 6 inches high, thin it down to two stalks. This will give a perfect stand. Every precaution should be used to secure uniform conditions in each plat; otherwise the experiment would be a waste of time, as the results would not mean anything.

Cultivation. — Ordinary cultivation should be followed, care being taken that none of the rows are unduly injured.

Taking notes. — Some breeders prefer to keep extensive descriptive notes for their own information, but for practical results, very little note-taking is necessary other than accurately to secure comparable yields. If the breeder is selecting for some particular quality, such as earliness, height of ear, angle of ear, and the like, he must take notes on these points. Also, taking a set of notes on each individual row furnishes the very best training possible in close observation; and as a man cannot know too much about the corn plant in order to be a successful breeder, it will usually pay him well to keep as complete a record as possible.

70. Harvesting. — When corn first ripens it contains 25 to 30 per cent of water, but it slowly dries out to about 15 per cent. In the breeding-plats some rows ripen and dry out sooner than others; hence, the weights will not be comparable until all are equally dry. For this reason it is best to leave the breeding-plats in the field for six to eight weeks after ripening, or until about December 1. Any very late-maturing or slow-maturing rows should be noted and discarded at harvest, as a type that will not mature well is undesirable.

A very good method of harvesting the plats is to divide a wagon box into two to four compartments. Husk a plat into each compartment. At the end of the rows, have a platform scale with a box large enough to hold the corn from one plat. Scoop the corn into this box, and as each plat is weighed, dump the corn at the end of the row, leaving the plat stake with each pile.

Leave the corn in these piles until all plats are husked, then mark the piles from high-yielding rows. A careful examination can now be made of these piles in order to note whether any seem immature, low in vitality, or otherwise undesirable. About one-fourth of the best plats should be noted, that is, 20 to 25 out of 100 piles. From these, seed for the general crop may be selected for the nest year.

The breeder still has one-half or more of the original ears from which the crop has grown. It is from these that he will build up his improved strains of corn.

71. The second year's work. — The best twenty or twenty-five original ears having been located, the remnants of these are again planted in separate rows the second year. The reason for so large a number of the remnants being again planted is because the degree of error may be so large — due to the fact that one season may favor a certain type — that we cannot determine exactly, the first year, just which are the best two or three for all seasons. When the second year's results are obtained, we may decide which to choose on the basis of two years' record. The seed from the best two or three rows may now be used as foundation stock for a select strain of corn. When the original ears are large, there will be quite a remnant left even after testing two years.

72. Continuation of breeding, several plans. — After the second year, a choice of several plans may be followed: —

1. Progeny of the best rows may be planted in a multiplying field for seed. Seed of this kind has given an increase of 9 bushels per acre (p. 92, Class II). It probably will not maintain its increased yield more than a few years without continued selection.

2. Ears may be selected from the best-yielding rows and the ear-row selection work continued. In this case it would be best to use some plan for preventing close breeding. Detassel every alternate ear-row plat having unrelated rows on each side. Save seed ears only from the best detasseled plats. This may be continued indefinitely, but it is probable that new ear-row plats should be started every two or three years for the purpose of securing new ears to be used as sire rows in the breeding block, thus giving a new stimulus through crossing. No work has yet been reported to show just what results are to be expected.

3. The original ear remnants may be used in a breeding block. Williams advocates this, using the best one or two ears for sires and detasseling the rest. The ear-row test is to be continued each year, securing ears from various sources, as the breeding-plat, general crop, or registered ears from other breeders. Each year the best remnants will be saved to be crossed the following year and passed into the multiplying plat the following season.

4. As the best-yielding ears may be hybrids of the

" elementary strains that nick well," plants from the best ear remnants could be used for inbreeding for the purpose of securing pure strains. There is more chance of securing strains that will cross well from these ears than when plants are taken at random. After pure strains had been secured that would cross to advantage as determined by experiment, these strains would be grown from year to year in isolated fields and used each year to produce first-generation hybrid seed.

SUMMARY

73. No fixed relation has been found between type and yield. The largest well-matured ears growing under normal conditions of soil and stand should be used for seed. Large ears with a medium depth of grain usually mature better than large ears with a very deep grain.

Mass selection and pedigree selection will ultimately give similar results when visible characters are chosen, as height of ear or shape of ear; but with invisible characters, as ability to yield, results are not so sure with mass as with pedigree selection, and at best they will come slowly

Pedigree selection involves the testing of each ear separately for yield, by planting a part. The remnants of best ears may be used directly as a foundation stock and the progeny may be continued in some system of ear-to-row selection. The very highest values will be secured by systematic crossing of the best strains.

CHAPTER X

RESULTS WITH HYBRIDIZATION

PERHAPS no cultivated plant has yielded so many interesting results of both practical and purely scientific value, as a result of hybridization, as the corn plant. Corn freely hybridizes, thus offering many opportunities for the selection of natural hybrids. The plant is so easily manipulated in artificial crossing, that almost any one may succeed with it who would fail with other crops. Not only have important results been secured bearing on the improvement of yield in quality of corn, but also interesting scientific data relating to the hereditary laws governing plants and animals, have also been secured with corn.

DEGREES OF RELATIONSHIP

74. When pollen of a maize plant falls from its own tassel on its own silk, this is called " inbreeding." When pollen from one variety is used to fertilize another variety, it is called " broad breeding." There are several intermediate grades of relationship, which may be summarized as follows : —

1. Inbreeding (pollen from own tassel).
2. Close breeding (pollen from sister plant, that is, plant from the same ear).
3. Narrow breeding (pollen from plants of the same variety).

FIG. 29. — Chart showing possible degrees of relationship between corn plants. (See text.)

4. Broad breeding (pollen from plants of a different variety).
5. Broad breeding (pollen from plants of a different group, as between flint and sweet corn).

XENIA

75. Cross a common dent corn on a sweet corn, using pollen from the former. When the ear is mature, instead

Fig. 30. — Method of covering tassel and ear for artificial pollination.

of the kernels being wrinkled and translucent as in sweet corn, a part of them will be smooth, resembling a dent corn. This is called xenia, or "the immediate effect of pollen on the endosperm." The effort of xenia, however, is limited to the kernel, there being no apparent effect on the cob or on the stalk.

In ordinary hybridization, only the germ is a hybrid and the endosperm surrounding is not affected. Xenia is accounted for by "double fecundation." In ordinary fertilization, only one of the two generative nuclei which are formed in the pollen tube is supposed to pass into the embryo sac and unite with the egg cell. In double fecundation both nuclei enter the egg sac, one fusing with the nucleus of the egg cell, and the other with the polar nuclei to form the embryo sac nucleus, the division of which gives rise to the endosperm. The endosperm would then be a hybrid and partake of the dominant characters of the male parent.

The fact that only a part of the kernels show xenia, means that double fertilization does not always take place. It is necessary that the germ cell be fertilized, but it appears at present that fertilization of the endosperm nucleus is incidental rather than necessary.

MENDEL'S LAWS

76. If crossed or hybrid kernels of dent and sweet corn be planted, they will produce ears having both dent and sweet corn kernels, with a ratio of three dent corn grains to one sweet corn grain. This is explained by assuming that the germ cells (pollen grains or ovaries) are either pure sweet corn or pure dent corn.

When a plant is grown from a hybrid seed, then, one-half the pollen grains will represent pure sweet corn, and one-half pure dent corn, and the same with the ovaries. While the plants may be hybrid, the sexual elements remain pure. In the process of fertilization a union producing a hybrid (sweet × dent or dent × sweet) will occur twice as often as a pure dent (dent × dent) or a pure sweet (sweet × sweet).

All the hybrid kernels resemble the dent corn kernels, so we apparently have three dent kernels to one sweet kernel on each self-fertilized hybrid ear. If the seed of one of these hybrid ears be sown, we apparently harvest three starchy ears to one sweet ear, as follows: —

Germ Cells	Character of Progeny
Starchy × starchy	Starchy
Starchy × sweet	Starchy
Sweet × starchy	Starchy
Sweet × sweet	Sweet

DOMINANT AND RECESSIVE CHARACTERS

77. In the above example, every hybrid ear has a starch grain and cannot be distinguished, on inspection, from a pure starchy type. Starchiness in this case is *dominant* over the sweet corn grain; or, in an ear that is a hybrid of dent corn and sweet corn, instead of the dent and sweet corn types blending, thus giving an intermediate, the dent completely dominates. In this case the sweet corn is *recessive.* By experiment this quality has been determined for many characters of maize, the following being typical examples: —

Red colors of cobs, stem, or husks dominant over green.
Starch endosperm dominant over sweet endosperm.
Yellow endosperm dominant over white endosperm.
Blue aleurone dominant over colorless aleurone.
Red pericarp dominant over colorless pericarp.
Podded kernels dominant over naked kernels.

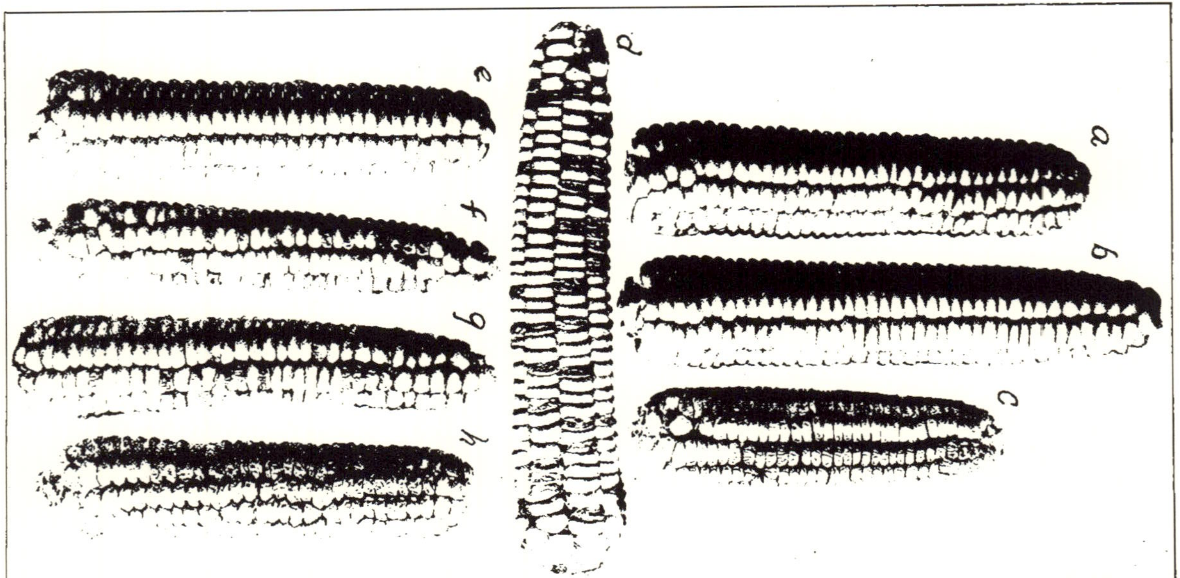

FIG. 31. — The segregation of types according to Mendel's law. *a*, is Rhode Island white cap, crossed with *c*, Crosby sweet corn. The ear *d* represents the first progeny of this cross, and shows no signs of sweet corn. *d*, a second generation (f_2) ear having one-fourth sweet kernels and three-fourths starchy. *e*, *f*, *g*, result from planting starchy kernels, one ear in three pure starchy. *h*, result of planting sweet kernels, pure sweet.

HYBRIDIZATION, EFFECT ON GROWTH

78. Two very distinct results follow cross-fertilization in maize: first, certain hereditary characters from both parents are bound up in the embryo, to be carried over into the next generation; second, a stimulus to vegetative growth is given, to be carried over into the hybrid generation. The carrying over of hereditary characters has already been discussed under the topics "Xenia" and "Mendel's Law," and the principal discussion here will deal with the second effect of hybridization.

SELF-FERTILIZATION

79. If a maize plant is self-fertilized (own pollen on own silk), and this process is repeated for two or three generations, and selected seeds are used, there may gradually be produced a "pure type." That is, the wide range of variation is decreased each year, until the progeny of the individual, being self-fertilized, are all of one type. For example, Shull self-fertilized corn plants of a white dent variety for five years and, as a result, secured strains that came true with 8, 10, or 12 rows, and so on up to 24-rowed ears. The pure strains differed also in other respects, but all the plants of a given pure strain were very similar. At the Nebraska Agricultural Experiment Station ten very distinct strains were isolated from Hogue's Yellow Dent, by inbreeding for three years.

The inbred strains also become dwarfish in size, have a high percentage of barren plants, and sometimes become entirely sterile.

The decrease in yield is illustrated by the following data from Shull. Two strains, designated as A and B, which had been inbred for five years and appeared to be pure strains, were compared with the original corn. The table also summarizes results with four strains of Leaming inbred by East, and a combination of several strains of Hogue's Dent inbred by Montgomery: —

TABLE XVI

Experimenter	Description of Seed	Yield per Acre of Original Strain Bushels	Yield per Acre of Inbred Strain. (Number of Years Inbred at Head of Column)				
			1st	2d	3d	4th	5th
Shull[1]	Pure strain A	79.4					14.2
	Pure strain B	79.4					12.1
East[2]		88	50.1	59.9	49.0	51.8	
Montgomery[3]		40.7			9.9		

[1] Ann. Rpt. Amer. Breeders Assoc., Vol. *VI*: 63–72. 1909.

[2] Conn. Agr. Exp. Sta., Bul. 168. 1911.

[3] Nebr. Agr. Exp. Sta. Ann. Rpt. 1912: 183.

It is evident from these data that the immediate effect of self-fertilization is to reduce the yield, the greatest reduction taking place the first year.

The amount of decrease seems to differ with varieties, or even with strains of the same variety. In general, inbreeding will decrease yield to about one-half the first year. Continuing the inbreeding will in some cases reduce the yield to one-fourth the original yield, while in other cases, inbred strains become entirely sterile. Present experiments indicate that the yield is reduced until the strain becomes a "pure strain," after which inbreeding has no further effect in decreasing yield.

Very often abnormal types appear in inbred strains.

FIG. 32.—The effect of three degrees of relationship in crossing is here illustrated. Nos. 3 and 4 are pure strains inbred for three years. No. 2 from a close fertilized seed-stock, the plants each year fertilized from sister plants from the same ear. No. 1 is from a seed-stock, cross fertilized for three years.

PURE STRAINS, OR BIOTYPES

80. Doctor Shull first presented the idea that a cornfield is to be considered more or less a mixture of pure types (biotypes) and that most of the plants in a field are more or less complex hybrids. By inbreeding, some of the original pure types might be isolated. In other words, a cornfield is a complex mixture of types, and inbreeding gives much the same result as does a chemical analysis with a complex compound—resolves it into its original elements.

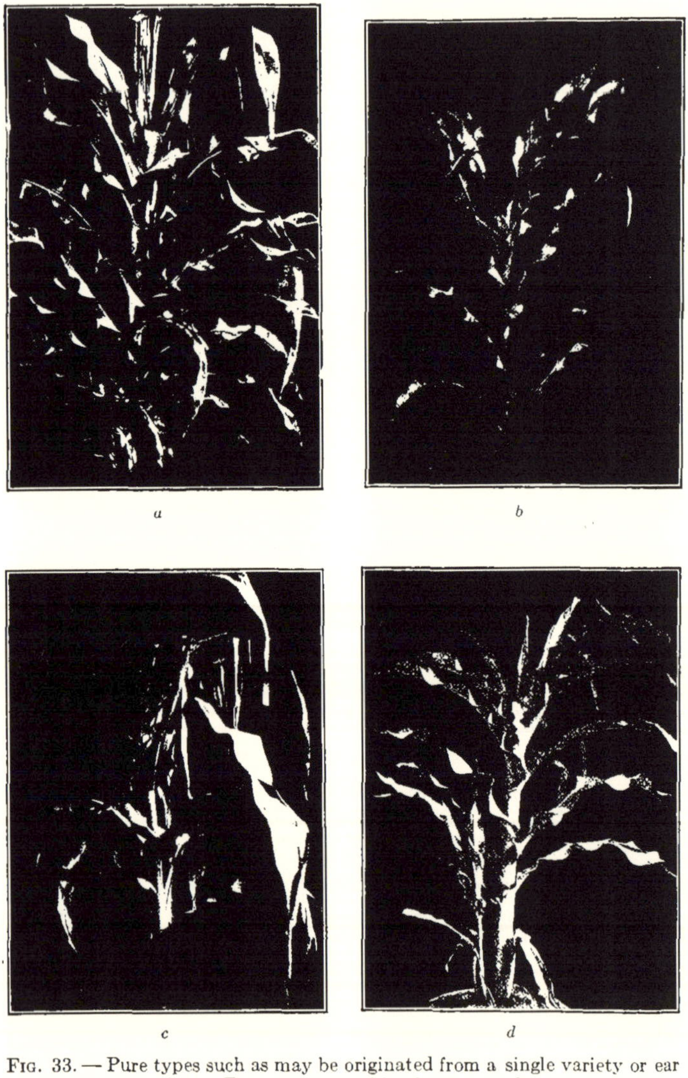

Fig. 33. — Pure types such as may be originated from a single variety or ear of corn by inbreeding. They are reduced in size, but each type comes true.

CROSSING BIOTYPES

81. However, these "elemental strains" are low in yield, but are stimulated to yield by hybridizing. The effect of hybridizing as a stimulus is shown in the following table: —

TABLE XVII

Experimenter	Description of Seed	Yield per Acre of Pure Strains Bushels	First-year Yield per Acre of Cross of Pure Strains Bushels	Second-year Yield per Acre of Progeny from Hybrid Seed Bushels
Shull . .	Pure type A Pure type B	14.2 12.1	79.8	69.5
East . .	Leaming No. 12 Leaming No. 9	35.4 23.3	110.2	98.4

The corn is at once restored to full vigor by hybridizing. The yield appears to decrease somewhat the second year, probably because a certain percentage of the plants have returned to a pure-type (homozygous) state.

Under natural conditions there is a certain percentage of both inbreeding and close breeding. This has led to the suggestion that means should be used to insure that all seed used is first-generation hybrid. This can be accomplished by alternating rows of pure strains and detasseling every alternate row, saving seed from the detasseled rows.

CROSSING VARIETIES

82. Several investigations have shown an increase from the use of first-generation hybrid seed, when two varieties have been crossed. The following table, summarized from

Morrow and Gardner's experiments in 1892 at the Illinois Agricultural Experiment Station, illustrates:[1]—

TABLE XVIII

Variety	Bushels of Air-dry Corn
Burr's White	64.2
Cranberry	61.6
Average	62.9
Cross	67.1
Burr's White	64.2
Helm's Improved	79.2
Average	71.7
Cross	73.1
Leaming	73.6
Golden Beauty	65.1
Average	69.3
Cross	86.2
Champion White Pearl	60.6
Leaming	73.6
Average	67.1
Cross	76.2
Burr's White	64.2
Edmonds	58.4
Average	61.3
Cross	78.5

The average yield of hybrids in the five tests was 9.7 bushels above the average of the parents and 4.5 bushels above the average of the highest parents. It is also shown, by this and other data, that certain crosses give

[1] Ill. Agr. Exp. Sta., Bul. 25. 1902.

greater increases than do others. Hartley has found that, while certain crosses give an increase, others give a decrease, and in some cases the cross is almost sterile. Probably those varieties that have been longest selected

FIG. 34. — A breeding plat where many tassels and ears are covered with paper bags, for artificial pollination.

as to type, and therefore are the nearest to a pure (homozygous) state, will respond most readily to crossing.

83. Shull has found that when a variety has been resolved, by inbreeding, into pure strains, certain of these pure strains when crossed give yields superior to the yield of the original corn, while other combinations give poor yields. He suggests that the maximum yields will be secured by first reducing a variety to its elemental strains, and then producing hybrid seed each year from only

those strains that give maximum results. This would necessitate maintaining the pure strains each in a separate field from year to year, and having another seed field where

FIG. 35. — Pure types as developed by inbreeding. No. 11 produces many tillers, and was reddish in color. No. 12 was free from tillers.

they would be planted in alternate rows. One strain would be detasseled in this seed patch, thus giving each year a stock of hybrid seed.

ISOLATING HIGH-YIELDING BIOTYPES

84. Evidence at present indicates that high-yielding ears, found by the ear-row method of testing, are in many cases natural hybrids of high-yielding biotypes. Thus by securing high-yielding ear remnants as foundation stock, they might be inbred until pure types were obtained. There is greater probability of securing biotypes that would combine to advantage from this stock than if a chance stock were used as a beginning.

SUMMARY

85. Fertilization is the result of the union of the contents of a pollen grain with the egg cell of an ovary. Xenia is the immediate effect of pollen in changing the character of the maize grain.

Mendel's law refers to the phenomenon of transmitting characters in toto, without blending, as in the case of dent and sweet corn when crossed.

Hybridization usually gives a decided stimulus to growth, while self-fertilization has the opposite effect. Continuous self-fertilization may reduce yield to one-fourth or less of the original yield, but the yield is fully restored in the first generation hybrids. A field of corn appears to be a miscellaneous mixture of biotypes, naturally not very productive, but stimulated to the highest degree of productivity by hybridizing. Certain biotypes hybridize to better advantage than do others.

References on xenia: —

WEBBER, H. J. (1900.) Xenia. U. S. Dept. Agr., Div. Veg. Physiol. and Path., Bul. 22.

GUIGNARD, L. (1901.) La Double Fecundation dans le Mais. Journ. Bot. [Paris], *15*: 1–14. No. 2.

References on inheritance in maize: —

CORRENS, C. (1901.) Bastarde zwischen Maisrassen mit Besonderer Berucksichtigung der Xenien. Bibliotheca Bot., *53*: 1-161.

LOCK, R. H. (1906.) Studies in Plant Breeding in the Tropics, 111. Ann. Roy. Bot. Gard. Peradeniya, *3*: 95–184.

EAST, E. M. A Note Concerning Inheritance in Sweet Corn. Science, N. S. *29*: 465–467.

1911. Inheritance in Maize. Conn. Agr. Exp. Sta., Bul. 167.

EMMERSON, R. A. (1911.) Genetic Correlations and Spurious Allelomorphism in Maize. 24th Ann. Rpt. Nebr. Agr. Exp. Sta.

References on crossing varieties: —

COLLINS, G. W. (1909.) The Importance of Broad Breeding in Corn. U. S. Dept. Agr., Bur. Plant Indus., Bul. 141.

(1910.) The Value of First-Generation Hybrids in Corn. U. S. Dept. Agr., Bur. Plant Indus., Bul. 191.

Increased Yields of Corn from Hybrid Seed. U. S. Dept. Agr. Yearbook *1910*: 319–328.

MORROW, G. E., and GARDNER, F. D. (1892.) Field Experiments with Corn. Ill. Agr. Exp. Sta., Bul. *25*: 173–203.

KELLERMAN, W. A., and SWINGLE, W. T. Ann. Rpt. Kans. Agr. Exp. Sta., No. *1*: 316–337, 1889; No. *2*: 288–355, 1890; and Kans. Agr. Exp. Sta., Bul. *27*: 139–158. (1891.)

EAST, E. M. Conn. Agr. Exp. Sta., Bul. 167. 1911.

HARTLEY, C. P., and associates. Cross-breeding Corn. U. S. Dept. Agr., Bur. Plant Indus., Bul. 218.

CHAPTER XI

ACCLIMATION AND YIELD

A BOTANICAL survey of the United States shows large, well-defined regions, each with a characteristic native flora.

86. Considering the great length of time that native vegetation has had in which to adjust itself, these various

FIG. 36. — Prairie vegetation in the "short grass" region. The natural vegetation indicates a very great difference in natural climate.

regions must indicate different environmental conditions, or else we should have a homogenous native vegetation throughout the country. Those regions covered with a forest vegetation must differ in environment (soil or climate) from a prairie region. There are various kinds of forest regions, as evergreen and deciduous; while in the prairies we have, along the Missouri River, a tall vegetation of grass and other plants, waist-high to a man, in

FIG. 37. — Prairie vegetation in humid region. Compare with Fig. 36. There must be quite a marked difference in the types of corn adapted to these two regions.

marked contrast to the " short grass " country three hundred miles westward.

Even within a State distinct floral zones can often be identified, as in Nebraska, for example, where six zones are recognized, each with a characteristic vegetation.

EFFECT OF ENVIRONMENT ON THE CORN PLANT

87. It has long been observed that each region would have a distinct type of corn plant. In northern regions the plant is leafy with the ear borne very low; in dry regions the plant is stocky, with a high proportion of ear and often with scant leaf; while in southern regions the plants are tall and have a low proportion of ear to stalk.

EFFECT OF PREVIOUS ENVIRONMENT ON YIELD

88. The marked effect of a change in environment on yield of grain has often been noted, the change usually decreasing the yield at first. At the Arkansas station,[1] 233 samples of corn were collected from various States and grown in comparison for two years.

In 155 trials with seven varieties, the highest yield was secured with seed grown between the thirty-fifth and thirty-eighth parallels of latitude, rather than either north or south of this region, this being the latitude of the Arkansas station.

TABLE XIX

Table showing the Yield of Corn from Seed of Different Sources at the Arkansas Agricultural Experiment Station

Names of Varieties	Number of Tests	Seed grown North of 38th Parallel of Latitude	Seed grown between 35th and 38th Parallels of Latitude	Seed grown South of 35th Parallel of Latitude
		Average for 2 Years	Average for 2 Years	Average for 2 Years
Leaming . . .	21	20.98	26.20	17.20
Golden Beauty .	20	32.81	45.775	50.475
Hickory King .	23	24.855	31.81	29.10
Golden Dent .	26	21.52	25.00	25.30
Champion White Pearl . . .	11	22.62	32.00	30.10
Early Mastodon	16	33.54	33.75	33.45
White Dent . .	38	24.175	34.695	34.775
Average Total	155	25.785	32.47	31.485

[1] Newman, C. L. (1899.) Ark. Agr. Exp. Sta., Bul. 59.

At the Nebraska Agricultural Experiment Station six leading varieties of corn were compared for two and three years, the seed in one case being native-grown and in the other from Iowa or Illinois. Results were as follows:—

TABLE XX

Table showing Yield of Corn from Acclimated Seed and Seed from Other Regions, at the Nebraska Agricultural Experiment Station

Name and Places of Origin		1903	1904	1905	Average	Difference
Silvermine	Nebraska		70.0	76.1	73.0	
	Illinois		65.1	63.4	64.2	8.8
Leaming	Nebraska		95.2	69.8	82.5	
	Illinois		76.6	72.3	74.4	8.1
Snowflake White	Nebraska	73.7	84.8	74.5	77.7	
	Iowa	68.7	72.8	67.1	69.5	8.2
Boone County White	Nebraska		76.2		76.2	
	Illinois		68.9		68.9	7.3
Early Yellow Rose	Nebraska	68.1	67.9	75.1	70.3	
	Iowa	62.1	76.9	63.5	67.5	2.8
Reid's Yellow Dent	Nebraska		83.8	64.2	73.7	
	Illinois		82.8	60.8	71.8	1.9
Average						6.2

In every case the native seed gave best results.

In another experiment conducted with farmers in western Nebraska, it was found that native-grown seed gave better results than seed grown in eastern Nebraska.[1] Rainfall in the western part of the State is very low, averaging about 18 inches annually, while the rainfall is about 30 inches in eastern Nebraska. To succeed in the West corn must be adapted to drought resistance.

[1] Nebr. Agr. Exp. Sta., Bul. 126. 1912.

TABLE XXI

TABLE SHOWING COMPARATIVE YIELD OF NATIVE- AND IMPORTED-SEED CORN IN WESTERN NEBRASKA IN BUSHELS PER ACRE

YEAR	VARIETIES NOT NATIVE (MOSTLY FROM EASTERN NEBRASKA)	NATIVE VARIETIES	DIFFERENCE
1908	24.1	30.5	6.4
1909	20.9	25.4	4.5
Average . . .	22.5	27.9	5.4

ADAPTATION OF THE SOIL

89. The climatic and soil requirements of corn have been stated in Section II. The climate cannot be controlled or modified in a marked degree, hence corn production is limited by climate to those regions where the natural rainfall, temperature, and like conditions are favorable to a profitable production.

The soil, however, is subject to treatment, and almost every soil can be brought to a high degree of productiveness by proper management. The subject of soil management is so fully treated in special texts on this topic, that it is not necessary to take up the matter in detail here.

From a study of corn soils as classified according to productiveness, it is apparent that a large proportion of the soil likely to be cultivated in corn may be grouped in two classes: first, soils that were once productive but are now more or less deplete by 50 to 200 years cropping; and second, soils that never were productive. In both cases the important factors to be modified can be grouped under three general heads, as follows: (1) organic matter, (2) mineral matter, (3) water.

CHAPTER XII

CROPPING SYSTEM IN RELATION TO MAINTAINING THE YIELD OF CORN

THE discussion so far, on the adaptation of soil for corn growing, brings out the fact that the constant growing of corn involves the development of a cropping system by which, with the least cost, the organic matter can be maintained and the most profitable use made of any fertilizing material that it may be necessary to add.

90. Cropping systems in the United States undergo evolution from the time when new land is opened up to the time when it reaches a permanent agricultural basis.

When new land is first brought under cultivation, grain farming is the general custom. Often a single crop is cultivated, as wheat in the Northwest. In a few years the single crop becomes unprofitable, due to the coming of insect pests or plant diseases, or to the decreasing availability of some mineral element in the soil. Then cultivated crops are introduced to alternate with the small grain.

In many regions of the Corn Belt, corn was continuously raised until it became necessary to introduce small grain culture. After a time, however, the continuous rotation of grain crops alone no longer gave paying crops. In general, this appears to be due to: —

1. Exhaustion of actual organic matter resulting in
   (*a*) Decrease in availability of some necessary element as phosphorus, or
   (*b*) Poor physical condition of the soil.

2. Exhaustion of some necessary element, usually lime, nitrogen, phosphorus, or potassium.

RESTORING PRODUCTION

91. When low production is due to the exhaustion of organic matter, then any cheap system of restoring that matter, as plowing under a green manure crop, will usually restore production in a measure. One effect of this decaying organic matter is the reaction on the minerals of the soil, thus increasing solubility, and restoring the physical condition. The *physical effect* is to make the soil more loamy in character by increasing granulation of clay, on the one hand, and on the other hand, in the case of sandy soils, binding the particles together. In this case no new supply of plant food is added to the soil, as the organic matter is grown on the land and only adds to the soil the carbon compounds taken from the air. Adding organic matter from an outside source, in addition to the above, also adds its own supply of elements.

When a certain element has been exhausted from the soil, that element may be added.

Nitrogen may be added in three ways: (1) by growing leguminous crops; (2) by adding organic matter from an outside source; (3) by adding nitrogen salts.

Phosphorus, potassium, and lime can be restored in two ways: (1) by adding organic matter from an outside source; (2) by adding salts of phosphorus, potassium, or lime.

Aside from a proper system of drainage where needed, the whole problem of devising a cropping system, including the application of fertilizers for maintenance or increase of production, is involved in the above statements.

Cropping systems may then be classed as: —

(1) Those that decrease productivity.
(2) Those that maintain productivity.
(3) Those that increase productivity (or in most cases merely restore it).

Experiments demonstrating the above cases have been made in a number of States where corn was used as one of the crops in the system.

MAINTAINING PRODUCTION

92. Results are reported from the Illinois station,[1] where corn has been grown in three systems of cropping, for 13 years in one case and for 29 years in the other.

TABLE XXII

Illinois Corn Yields where Three Systems of Cropping are Compared. Average Yield for Last Three Years

Crop Years	Crop System	13-year Experiments Bushels	29-year Experiments Bushels
1905–6–7 . .	Corn every year	35	27
1903–5–7 . .	Corn and oats	62	46
1901–4–7 . .	Corn, oats, clover	66	58

The land on which these experiments were conducted originally produced more than 70 bushels per acre. There has been some decrease in yield in all cases, but less decrease where rotation was practiced. Yield cannot be maintained by rotation alone *where the crops are removed.*

In a second series of plots a corn-oats-clover rotation was practiced, where all was returned to the land except the grain and clover seed harvested. In one case, the straw,

[1] Ill. Agr. Exp. Sta., Bul. *125* : 324. 1908.

cornstalks, and clover were all plowed under, and this system was designated as "grain farming" since no live stock to produce manure was needed.

FIG. 38. — Good land, continuously cropped with grain, until it is in an unproductive state.

In a second series, designated "live-stock farming," the crops have been removed but equivalent manure returned.

CROP YEARS	SPECIAL TREATMENT	GRAIN FARMING LEGUMES[1] BUSHELS	LIVE-STOCK FARMING MANURE[2] BUSHELS
1905-6-7 . .	None	69	81
1905-6-7 . .	Lime	72	85
1905-6-7 . .	Lime, phosphorus	90	93
1905-6-7 . .	Lime, phosphorus, potassium	94	96

[1] Legume catch-crops and crop residues.

[2] Manure applied in proportion to previous crop yields.

Growing legumes and returning all residues has maintained yield, and when in the form of manure has increased yield. When additional minerals have been added, the crop production has been actually increased about 20 per cent.

At the Indiana station[1] five cropping systems have been compared for 20 years, with and without commercial fertilizers. Without giving details, the following table shows clearly enough the comparative effect of different cropping systems on the maintenance of production.

TABLE XXIII

INDIANA EXPERIMENTS, COMPARING CORN YIELDS AT BEGINNING AND END OF TWENTY-YEAR PERIOD IN DIFFERENT ROTATIONS

Treatment	Cropping Systems and Yields in Bushels per Acre				
	I Continuous Corn	II Corn and Wheat	III Corn, Oats, Wheat-Clover	IV Corn, Oats, Wheat, Clover-Grass	V Corn-roots, Oats, Wheat, Clover-Grass
	Yields in 1889 when the Experiments were Begun				
Unfertilized .	61.1	50.0	54.6	54.2	58.4
Fertilized .	62.1	49.3	54.8	56.4	58.1
	Yields in 1909 after 20 Years' Cropping				
Unfertilized .	26.0	25.4	47.8	35.5	61.1
Fertilized .	39.9	47.3	59.1	65.5	73.1
	Difference between 1889 and 1909 Yields, showing Effects of Rotations				
Unfertilized .	−35.1	−24.6	− 6.8	− 18.7	+ 2.7
Fertilized .	−22.2	−2.0	+ 4.3	+ 9.1	+ 15.0

[1] Ind. Agr. Exp. Sta., Circ. 25. 1911.

The Indiana results confirm the Illinois experiments, showing: (*a*) a rapid decrease for continuous grain culture; (*b*) a maintenance of yield for longer period when a rotation including legumes and grass is included; (*c*) an

Fig. 39. — Compare with Fig. 38. The same kind of land near by, but properly managed to maintain productivity. (Minn. Exp. Sta.)

actual increase in productivity when fertilizer is added. *However, fertilizer did not maintain yield in a grain rotation.*

ROTATIONS FOR CORN GROWING

93. The above tables suggest the type of rotations and fertilizer treatment for the Corn Belt. Other stations have suggested rotations including corn, as follows:[1] —

"A rotation for dairy farms recommended by the New Jersey station consists of (1) field corn, seed to crimson clover in July or August; (2) crimson clover followed by

[1] U. S. Dept. Agr., Farmers' Bul. *144*: 11.

fodder corn, land seeded to winter rye; (3) rye fodder, followed by oats and peas, seeded to red clover and timothy; (4) hay. [Crimson clover is not hardy north of New Jersey. — AUTHOR.]

"A three-year rotation for the South, recommended by the Louisiana station, is (1) corn; (2) oats, followed by cowpeas; and (3) cotton.

"At the Delaware station a good rotation for a poor soil in bad condition was (1) sweet corn, crimson clover; (2) cowpeas, winter oats; and (3) red clover. A fertilizer was applied. The results reported indicate that it is better to have crops growing continuously up in the land, than to have it lying idle during a part of the growing season."

Each farmer must work out the rotation system best adapted to his own situation, but the general lines to follow are indicated in the foregoing discussion.

CHAPTER XIII

ORGANIC MATTER FOR CORN LAND

Organic matter has several important functions in the soil: (1) As a direct source of food supply. Decaying vegetable and animal matter contains all the essential food elements of plants. (2) As a means of freeing unavailable plant food elements in the soil. The organic acids given off by decaying organic matter act directly on the elements of the soil, in bringing them into solution. (3) The physical condition of the soil is affected in a remarkable degree by the presence of even a small percentage of organic matter. Note the effect on a clay soil when a few loads of manure are applied to an acre of land. The organic matter improves the granulation and increases the water-holding capacity to some extent. Aëration is also improved. (4) A very important effect is to improve the soil as a medium for the growth of soil bacteria and fungi, which in turn become a source of organic matter to the soil. (5) Nitrogen-fixing bacteria are favored by abundant organic matter, if sufficient lime be present.

Considering the fact that corn, in common with all cereals, must be grown without the extensive use of commercial forms of fertilizer, *maintaining the supply of organic matter in the soil becomes the most important single consideration in extensive corn-growing regions.*

94. Good corn soils are rich in organic matter. Two of the best corn soils in the Central States are Miami black clay loam and Wabash silt loam, the organic matter of which, according to Lyon and Fippin,[1] is as follows: —

	Soil 0–7 Inches Per Cent Organic Matter	Subsoil 7–40 Inches Per Cent Organic Matter
Miami black clay loam Av. 12 samples . . .	5.9	2.50
Wabash silt loam Av. 11 samples . . .	3.3	1.30

Of all the cereals corn is the best crop to grow first, after a heavy application of manure or the plowing under of organic matter, as a clover sod or green manure crop. It is well adapted to utilize the rather large store of nitrogen likely to become available at such time. Because corn does well following the plowing under of coarse organic matter, it is sometimes called a " coarse feeder "; while wheat, requiring a more advanced stage of decomposition, is termed a " delicate feeder."

FARMYARD MANURE FOR CORN

95. It has been demonstrated that lime, under certain conditions, applied to the land gave profitable increases; in certain other cases commercial fertilizers have been profitable; but farmyard manure, wherever used, has usually given profitable returns. It appears at present, however, that for a large share of the corn-growing area farmers are not justified in keeping sufficient live stock in their farming systems to depend on manure as the principal means of maintaining production. It must be

[1] Lyon and Fippin. Soils, p. 125.

TABLE XXIV

RESULTS FROM PENNSYLVANIA AND OHIO EXPERIMENT STATIONS SHOWING COMPARATIVE RESULTS WITH COMPLETE FERTILIZERS AND FARMYARD MANURE ON THE YIELD OF CORN. FINANCIAL GAIN FOR THE WHOLE ROTATION

Station	Rotation	Yield of Corn (Bushels per Acre)			Financial Statement for One Rotation—All Crops				
					Fertilizer			Manure	Manure
		No Fertilizer	Complete Fertilizer	Farmyard Manure	Cost per Acre	Value of Increase	Net Increase	Value of Increase	Value per Ton
Pennsylvania 25-yr. average Bul. 190	Corn-oats Wheat-clover	42.1	56.90	(12 tons) 57.5	$21.44	$28.08	$6.64	$25.96	$2.16
Ohio 16-yr. average Circ. 104, p. 10	Corn-oats Wheat-hay Hay	30.4	44.63	(8 tons) 44.03 (16 tons) 51.11	$19.29	$29.35	$10.06	$30.00	$2.50
Ohio . . . . 13-yr. average . Circ. 104, p. 17	Corn-wheat Clover	33.8	43.74	(8 tons) 54.66	$4.88	$10.36	$5.48	$21.80	$2.72
Average .		35.4	48.42	51.82	$15.20	$22.60	$7.39	$25.92	$2.46

supplemented by plowing under organic matter, especially green crops, containing enough legume crops to maintain the nitrogen supply.

Perhaps the best comparative idea of the value of farm-yard manure and fertilizers can be obtained by examining certain data secured at the Pennsylvania station in a twenty-five-year test, two experiments at the Ohio station — one for sixteen years and the other for thirteen years — a nine-year test at the Indiana station, a thirteen-year record at the Illinois station, and a single corn crop after timothy at the Cornell station.

The foregoing table is a summary of the data secured at the Pennsylvania and Ohio stations. These are the best continuous records that we have in the older portion of the United States, where the use of manure and fertilizers is now becoming a matter of importance.

The summary shows that land yielding, under a good rotation, an average of 35.4 bushels of corn per acre has been maintained at an average of 48.4 bushels for the corn crop by an average expenditure of $15.20 for commercial fertilizers (where a complete fertilizer was used) for each course of the rotation (average of four years). These complete fertilizers were fairly well mixed to meet the needs of the soils in each case. An average application of 11 tons of manure every four years has maintained the yield of corn at 51.82 bushels.

The second part of the table shows the average financial returns for all crops grown during the course of rotation. Eleven tons of manure shows a better average return than $15.20 worth of commercial fertilizer, and an average return of $2.46 per ton for the manure. The Illinois station received a return of $1.60 per ton and the Indiana station $1.50 per ton for manure. Both the latter stations are on newer land, where as large increases are not yet to be expected as on old cultivated land.

Again, the Ohio station[1] has shown that the value of manure may be increased by adding and composting a small amount of mineral fertilizer with it. The following table summarizes the data, showing a marked increase in the value of the manure where treated with mineral fertilizer.

TABLE XXV

VALUE OF BARNYARD MANURE TREATED WITH 40 POUNDS PER TON OF DIFFERENT MINERALS. APPLIED TO CROPS IN A THREE-YEAR ROTATION OF CORN, WHEAT, AND HAY AT THE RATE OF 8 TONS PER ACRE. AVERAGE FOR FOURTEEN YEARS

PLATS NO.	AMENDMENT USED	COST OF 40 POUNDS CENTS	TOTAL VALUE OF INCREASE IN ONE ROTATION	NET VALUE OF INCREASE PER TON OF MANURE
2–3 . . .	Floats	18	$33.51	$4.01
5–6 . . .	Acid phosphate	30	38.08	4.46
8–9 . . .	Kainit	34	29.28	3.32
12–13 . .	Gypsum	12	27.41	3.30
15–16 . .	Untreated	0	23.44	2.93

In this case, the mineral was mixed with the manure as it was removed from the barns. A part was applied directly to the land in each case, and a part allowed to decay in the yards for about three months. The former method seemed to be the better.

SUMMARY

96. In the foregoing discussion on the theory of applying fertilizers and manures for raising cereals, it appears that the permanent maintenance of the soil in a productive

[1] Ohio Agr. Exp. Sta., Circ. *120* : 112. 1912.

state is the most fundamental problem in production. For cereal culture, the soil must be maintained at the lowest possible cost. The four principal elements to be given attention in most soils are (1) nitrogen, (2) phosphorus, (3) potash, (4) lime. In addition, active organic matter must be present.

These conditions are met in the most practical way by:

(1) A rotation in which legumes furnish a large share of the nitrogen used by other crops in the rotation.

(2) Where manure is not available, practically all organic matter grown on the land, with the exception of threshed grain, should be plowed under.

(3) In order to maintain the full supply of organic matter and nitrogen, it may be necessary to plow under the entire legume crop without harvesting.

(4) Where live stock is kept, all manure made by feeding produce should be returned to the land in relatively light dressings.

(5) The constant removal of grain will gradually reduce the phosphorus and potassium. This must be returned as commerical fertilizer. A part at least can be mixed with the manure and applied in this way.

(6) Where fertilizer mixtures are applied to land, careful regard should be given to the needs of the land, and the fertilizer should be mixed to meet the particular needs in each case.

(7) Where lime is required, it should be applied once every four to six years, the amount being determined by the needs of the land.

CHAPTER XIV

MINERAL MATTER FOR CORN LAND

As pointed out heretofore (p. 42), about 1 per cent of the green weight or 5 per cent of the dry weight of corn is ash or mineral matter, taken directly from the soil. For the production of 50 bushels of corn the mineral ash found in composition would be as follows: —

TABLE XXVI

MINERAL AND NITROGEN REQUIREMENTS OF A 50-BUSHEL CORN CROP[1]

	NITROGEN	PHOSPHORUS	POTASSIUM	MAGNESIUM	CALCIUM	SULFUR	TOTAL MINERAL	TOTAL ASH
Ears (3500 lb.) . .	50.0	8.50	9.50	3.85	.7	.14	24.29	43.4
Stalks (3000 lb.) .	24.0	3.00	26.00	4.80	10.4	3.00	64.40	69.5
Total .	74.0	11.50	35.50	8.65	11.1	3.14	88.69	112.9

97. Soils in regard to mineral supplies may be classed as:

1. Soils in which sufficient mineral matter in available form is present.

2. Soils in which sufficient minerals are present, but one or more of those minerals are unavailable or available in very limited amounts.

3. Soils in which one or more minerals are so deficient that even with good soil management a sufficient amount cannot be made available for a crop, the total supply being sufficient for only a few crops.

[1] HOPKINS, C. G. Soil Fertility, pp. 154 and 603.

The first class included most of the present Corn Belt States when the land was first broken up. Large crops were grown without soil amendments, but to-day the yield is limited on most of these soils by the lack in available form of one or more mineral elements.

In the second class, chemical analysis may show the presence of enough minerals and nitrogen for fifty to one hundred crops, and yet the crop be limited, as the minerals may become available only very slowly. This class includes a large share of the above-mentioned soils that have been farmed fifty to one hundred years or more. Decreased availability of minerals may be due to several causes, as deficiency in bases such as lime or magnesium, or more often insufficient organic matter in a state of active decomposition. The addition of lime or organic matter, or both, is the evident remedy in such cases.

In other cases there is no practical way of making available sufficient mineral elements for maximum crops, and mineral fertilizers must be added. In many soils there are other inhibiting factors to plant growth, even when mineral elements are abundant. This is especially true on poorly drained soils where toxic organic compounds are developed.

In class 3 are included many of the sandy soils and soils subject to leaching, erosion, or derived from rocks that originally lack some mineral in composition. It is doubtful whether corn culture can ever be profitably developed on land of this class. An example is the sandy soil of Long Island, where most of the mineral must be supplied. Often a ton or more per acre of high-grade fertilizer is used. On such land only crops returning a large gross income per acre, as potatoes, cabbage, or truck, can be grown with profit.

98. Hopkins[1] believes it fair to "assume for a rough estimation that the equivalent of 2 per cent of the nitrogen, 1 per cent of the phosphorus, and $\frac{1}{4}$ of 1 per cent of the total potassium contained in the surface soil can be made available during one season by practical methods of farming."

The above statement is borne out by results in many of the prairie soils of Illinois. The amount of nitrogen, phosphorus, and potassium in the surface, and the amount available annually on the above basis, is shown by Hopkins to be as follows: —

TABLE XXVII

FERTILITY IN ILLINOIS SOILS AND AMOUNT ANNUALLY AVAILABLE IN POUNDS PER ACRE (ROUGHLY ESTIMATED)

Soil Type No.		Average per Acre in Surface Soil[2] (0-6⅔ in.)			Annually Available[3]		
		Total Nitrogen	Total Phosphorus	Total Potassium	Nitrogen	Phosphorus	Potassium
330	Gray silt loam	2,880	840	24,940	58	8	62
426	Brown silt loam	4,370	1,170	32,240	87	12	81
520	Black clay loam	6,760	1,690	29,770	135	17	74
635	Yellow silt loam	2,390	850	37,180	48	9	93
1401	Deep peat	34,880	1,960	2,930	9	20	7
Amount required for 50-bushel corn crop					74	11.5	35.5

From the above typical examples, it appears that many of these soils do not meet the requirements of a 50-bushel corn crop in all the three elements considered.

[1] HOPKINS, C. G., *l.c.*, p. 107.

[2] *Ibid.*, p. 82.

[3] *Ibid.*, p. 110.

The problem of production on soils of this class is to increase availability through use of manures and organic matter, but in many cases the addition of some mineral supplement is now necessary.

FERTILIZERS FOR CORN

99. Theory of fertilizer dosage. — If a perfectly sterile sand were used as a medium for growing crops, and it were desired to add fertilizing material, the logical method would be to ascertain the relative amount of mineral constituents in the plant under culture and add the minerals in the same relative proportion. For example, the three principal mineral constituents in the corn crop is shown in Table XXVI to equal, in a 50-bushel corn crop, 74.0 pounds of nitrogen, 11.5 pounds of phosphorus (or 26.3 phosphoric acid), and 35.5 pounds of potassium (or 42.6 potash); or the ratio would be about 6 : 2 : 3 for nitrogen, phosphoric acid, and potash.

If the amount of fertilizer applied were to equal the expected crop, then for a 50-bushel corn crop we should apply about the following formula:

74 lb. nitrogen = 400 lb. sodium nitrate
11.5 lb. phosphorus = 190 lb. acid phosphate
35.5 lb. potassium = 85 lb. muriate of potash

Fertilizer for corn would not, however, be applied to a sterile soil, but to a soil usually containing enough minerals and nitrogen in an unavailable state for fifty to one thousand crops. Organic matter and lime, and thorough tillage, assist in making minerals available; but after all reasonably good treatment has been given, some one or more elements may be found available only in such small amount that the crop is limited.

The fertilizer applied should be planned to supply the needed element or elements, rather than all elements. Also, a certain element may be deficient a part of the season but more plentiful at some other period. This is true of nitrogen, which is often deficient in early spring, especially on heavy clay soils, but may be more abundant by mid-summer.

The Ohio Agricultural Experiment Station reports an experiment in which fertilizers were applied in arbitrary quantities in comparison with plats on which "the three fertilizing elements, nitrogen, phosphorus, and potassium, are given in approximately the same ratio to each other in which they are found in the plant." [1]

TABLE XXVIII

FERTILIZER TESTS WITH CONTINUOUS CORN CULTURE AT THE OHIO AGRICULTURAL EXPERIMENT STATION. AVERAGE FOR SIXTEEN YEARS, 1894–1909

Plot No.	Fertilizing Materials Pounds per Acre		Yield		Increase	
			Grain Bushels	Stover Pounds	Grain Bushels	Stover Pounds
1	None		22.22	1441		
2	Acid phos. 160 Mur. potash 100 Nitrate soda 160	arbitrary quantity	42.71	2326	22.08	949
3	Acid. phos. 60 Mur. potash 30 Nitrate soda 160	ratio in corn plant	34.95	1946	15.90	634
4	None		17.46	1248		

The ratio between the elements in the two mixtures and that required by the plant is shown in the following statement: —

[1] Ohio Agr. Exp. Sta., Circ. No. *104*: 3. 1910.

	Nitrogen Pounds	Phosphoric Acid Pounds	Potash Pounds
Arbitrary mixture	24	24	50
Ratio	6	6	12
Natural proportion	24	8	15
Ratio	6	2	4
Elements required for 40 bushels corn	59	21.2	34.0
Ratio	6	2	3

The arbitrary mixture had approximately three times the phosphoric acid and potash in proportion to nitrogen that the natural proportion showed.

In this case the arbitrary mixture gave the best results, as the crop was able to obtain nitrogen from the soil to balance the fertilizer applied. The point is well illustrated in a second experiment in which the fertilizer mixtures were compared. The fertilizer was applied to corn in a three-year rotation of clover, corn, and wheat. A part of the benefit of the fertilizer went to the wheat and clover. Results with all three crops are given on the following page.

Plot 19 received a smaller application of fertilizer at less cost, yet it contained twice as much phosphorus, which seems to be the one element that this soil most required.

The above table emphasizes that the corn grower should handle nitrogen, phosphorus, and potassium more or less independently, adjusting his fertilizer application to the needs of the soil. The ready mixed fertilizer will not usually be as profitable as the fertilizer mixed especially for the case concerned.

TABLE XXIX

SHOWING FERTILIZERS APPLIED IN CERTAIN EXPERIMENTS AT THE OHIO AGRICULTURAL EXPERIMENT STATION IN A THREE-YEARS ROTATION OF CLOVER, CORN, AND WHEAT[1]

Plat	Fertilizer Pounds per Acre	Pounds of Elements applied per Acre: Nitrogen	P_2O_5	K_2O	Cost per Acre
17	None				
18	Nitrate soda 160; Acid phos. 80; Mur. potash 80 . . .	24	12.8	40	$7.45
	Ratio	7	4	11	
19	Tankage 100; Acid phos. 80; Mur. potash 10 . . .	7	23	5	$2.30
20	None				
	Approximate ratio in plant (See table.)	7	2.5	4	

Plat	Corn, Twelve Crops: Grain Bushels	Corn, Twelve Crops: Stover Pounds	Average Annual Increase per Acre: Corn Grain Bushels	Twelve Crops Stover Pounds	Wheat Grain Bushels	Twelve Crops Straw Pounds	Hay Pounds	Cost of Treatment	Value of Increase per Acre
17	36.55	2303							
18	43.12	2587	9.25	471	2.83	309	599	$7.45	$9.37
19	44.37	2456	10.50	348	4.07	510	718	$2.30	$11.36
20	34.09	2025							

[1] pp. 17–18.

FERTILIZER MIXTURES FOR CORN

100. To mix the fertilizer so as to suit the requirements of the particular soil and crop, is the ideal way. As a basis for use when the fertilizer requirements are not known, general experience indicates that a formula of about 3–8–5 will most often be satisfactory. The Maine Agricultural Experiment Station[1] suggests the following formula: —

TABLE XXX

FORMULAS FOR FERTILIZERS SUGGESTED BY THE MAINE AGRICULTURAL EXPERIMENT STATION

Crop and Fertilizing Material	Weight Used Pounds	Nitrogen Pounds	Phosphoric Acid Available Pounds	Potash Pounds
Corn on sod land, or in conjunction with farm manure.				
Nitrate of soda	100	16		
Acid phosphate	400		52	
Muriate of potash	150			75
Total	650	16	52	75
Percentage composition		*2.5*	*8.0*	*11.5*
Nitrate of soda	100	16		
Screened tankage	200	11	15	
Acid phosphate	300		39	
Muriate of potash	150			75
Total	750	27	54	75
Percentage composition		*3.6*	*7.2*	*10*
Nitrate of soda	100	16		
Cottonseed meal	200	14		4
Acid phosphate	400		52	
Muriate of potash	150			75
Total	850	30	52	79
Percentage composition		*3.5*	*6.1*	*9.3*

[1] Maine Agr. Exp. Sta., Bul. 107. 1904.

Director Charles D. Woods, who prepared the above formulas, makes the following statement in connection therewith: "Corn is a crop that uses a large amount of nitrogen. It is usually grown upon sod land or with farm manure, or both. Indeed, it is doubtful if, under ordinary conditions, it would prove a profitable crop to be grown on somewhat exhausted soil with commercial fertilizers alone. . . . The first formula contains only about one-sixth of the nitrogen needed to grow the crop. With a good sod and especially with a liberal dressing of farm manure, that will be all that is needed. The second and third formulas carry more nitrogen. . . ."

101. The New York (Geneva) Agricultural Experiment Station[1] suggests the following formulas for corn: —

TABLE XXXI

Formula	Pounds of Different Constituents for One Acre		
	Principal Source of Nitrogen	Principal Source of Phosphoric Acid	Principal Source of Potash
1	60 to 100 lb. nitrate of soda	350 to 700 lb. bone meal	60 to 120 lb. muriate of potash
2	50 to 100 lb. sulphate of ammonia	250 to 500 lb. dissolved bone	60 to 120 lb. sulphate of potash
3	100 to 200 lb. dried blood	300 to 600 lb. dissolved rock	250 to 500 lb. kainit
4	3000 to 4000 lb. stable manure		600 to 1200 lb. wood ashes
Pounds per acre . .	Nitrogen 10 to 20	Available phosphoric acid 35 to 70	Potash 30 to 60
Percentage	2	7	6

[1] N. Y. (Geneva) Agr. Exp. Sta., Geneva, N.Y., 14th Rept.

102. As soils are continuously cropped, progressive changes take place. A suggested method of adapting the fertilizer to conditions is given by C. E. Thorne of the Ohio Agricultural Experiment Station, as indicated by experience on rather poor glacial soil at that station:[1] —

TABLE XXXII

FERTILIZERS SUGGESTED FOR DIFFERENT CONDITIONS

Conditions	Percentage Composition		
	Ammonia	Phosphoric Acid	Potash
For crops immediately following clover	1	13	2
For crops one or two years after clover	3	12	3
For crops two or three years after clover	4	12	4
For crops on exhausted soils .	6	12	6

WHEN IT PAYS TO FERTILIZE FOR CORN

103. The gross income per acre from cereal crops is low, and their extensive culture can be carried on only where the soil naturally furnishes most of the mineral elements without excessive cost. In the past, cereal culture has largely followed the opening up of new lands, while it has declined on old soils when extensive use of commercial fertilizers has become necessary.

From the foregoing discussion it seems that the use of mineral fertilizers for corn can be applied at a profit only as a supplement to soils already well supplied with available minerals. In many cases when a single mineral

[1] Ohio Agr. Exp. Sta., Bul. 141. 1903.

element is lacking in an available from, this element may often be directly supplied at a profit; but ordinarily, in order to obtain the highest availability from the minerals, fertilizers must be used in connection with barnyard manures, and in a properly balanced crop rotation where most of the nitrogen is supplied by legumes and the soil is kept well supplied with decaying organic matter.

A review of the experimental evidence regarding the use of commercial fertilizers for corn seems to justify the following principles.

1. It seldom pays to use mineral fertilizers alone on land in a low state of fertility or on land that would not produce more than 20 bushels of corn per acre under favorable conditions.[1]

2. Even on good land it seldom pays to apply mineral fertilizer alone directly to the corn crop.[2]

3. It seldom pays to use fertilizers where corn is grown continuously or where it is rotated with grain crops only. Under such conditions, according to the Ohio station, only 60 per cent of the fertilizer is recovered in the crop.[3]

4. Commercial fertilizer pays, as a rule, only when used in connection with a rotation where manure or a legume crop, or both, are plowed under.[4] In this case it is usually best to apply the fertilizer to the sod land, or, when wheat is grown in the rotation, a part may be applied to the wheat. The above expecially applies to phosphates and potash. Sodium nitrate is a partial exception to the above general rule, as it is sometimes applied with profit to the growing corn.

[1] U. S. Dept. Agr., Farmers' Bul. *144*: 10; Farmers' Bul. *414*: 12. 1910. R. I. Agr. Exp. Sta., Bul. *113*: 113. 1906.

[2] Ind. Agr. Exp. Sta., Bul. *77*: 32. 1899.

[3] Ohio Agr. Exp. Sta., Bul. *110*: 68. 1899.

[4] U. S. Dept. Agr., Farmers' Bul. *144*: 10. 1901.

Special cases: There are exceptions to the above rules, a striking example of which are certain rich muck lands in Illinois, well supplied with all elements except potassium, where an application of potassium salts pays large returns.[1]

It is not to be inferred that fertilizers do not afford a stimulus and give increased production, for they do; but the gross income from an acre of corn is not sufficiently increased to pay the cost of fertilizer, except in certain cases when used in connection with manure and legumes. This makes it apparent that profitable corn growing must be carried on as a part of a general farming scheme in which the soil fertility is principally maintained by the use of green manures or barnyard manure, which may be supplemented in a limited way with commercial fertilizer.

NITROGEN

104. A large or excessive supply of available nitrogen is not considered favorable for most of the cereals, as wheat, oats, or barley; the effect being to produce an excessive growth of straw, and often a decreased yield of grain. Corn, however, is not so affected, and is most productive on heavily manured land or on newly drained alluvial or swamp lands where the available nitrogen is so abundant that wheat or oats would "run to straw" and produce little or no grain. In fact, a well-manured clover sod where available nitrogen is in greater excess than any other necessary element is ideal corn land.

A large supply of nitrogen has sometimes been found a disadvantage early in the season, as it may stimulate a growth of plant too large to be adequately maintained during the remainder of the season. For example, the

[1] Hopkins, C. G. Soil Fertility and Permanent Agriculture, p. 471.

"Williamson"[1] method of corn culture advocates the withholding of soluble nitrate fertilizer until the plants are six to eight weeks old, thus tending to retard stalk growth but to give the needed stimulus at the time when ears are forming.

West of the Missouri River, where the soil is loose and nitrification begins early in the season, it often happens that on very fertile soil a vigorous spring growth is stimulated, and later, if the season proves unusually dry, the growth cannot be sustained; and such fields suffer more than do fields in a less fertile condition. On the other hand, with abundant water supply those fields would have been more productive.

LIME

105. Lime is an essential element required by plants. It is not commonly applied as a fertilizer, as only about 12 pounds of lime are required by a 50-bushel corn crop, and most soils are abundantly supplied in so far as having sufficient lime for plant growth is concerned.

The most important use of lime is as a soil amendment, when it assists in several ways in making the soil more favorable for plant growth: —

1. Acid in the soil is neutralized.
2. Potash and phosphate in the soil are made more readily available.
3. Organic matter decays more rapidly and the organic nitrogen and minerals become available to plants in less time.
4. The soil is made a more favorable medium for beneficial bacteria forms.

[1] The Williamson Plan. S. C. Agr. Exp. Sta., Bul. 135. 1908.

5. The mechanical condition of heavy clay soils is improved.

According to Bulletin 64 of the Bureau of Soils, United States Department of Agriculture, one hundred sixty-eight experiments with lime for corn have been reported by experiment stations. The average increase reported is 3.2 bushels per acre at a cost of $8.91 for the lime. For corn soils in general liming would not pay, but, on the other hand, certain experiments show large profits from the use of lime.

The Tennessee station reports an increased yield, at less cost per bushel, than for any of a number of mineral fertilizers tried in comparison, as shown in the following table: —

TABLE XXXIII

Fertilizers with Hickory King Corn, 1901–1902

Fertilizer Used	Amount	Cost of Fertilizer per Acre	Yield per Acre of Grain Bushels	Increase Due to Fertilizer Bushels	Cost per Bushel of Increase
No fertilizer			41.94		
Farmyard manure	8 tons	$3.20	48.71	6.77	$0.47
Lime	25 bushels	1.50	49.22	7.28	.10
Nitrate of soda Acid phosphate Muriate of potash	100 pounds 150 pounds 5 pounds	4.00	43.97	2.03	1.74

At the Ohio station[1] the addition of lime increased the yield of corn 10 bushels per acre, or about 30 per cent, both

[1] Ohio Agr. Exp. Sta., Bul. *159* : 173.

on plats where lime was used alone and where it was used in connection with other fertilizers. In commenting on results with lime, Director Thorne says: —

"Taking the results as a whole, it would seem that the lime has performed two distinct offices in this test: in the first place, it has increased the yield by an average of about 10 bushels per acre, or 30 per cent of the unfertilized yield. This it must have done in one or both of two ways; either it has furnished a needed element of plant food to the growing crop, or else it has rendered the plant food already in the soil more available, either by direct chemical action of the lime itself on the soil stores of nitrogen, phosphorus, and potassium, or by opening up the soil and giving the air, water, and frost a better opportunity to reach these stores and prepare them for plant nutrition.

"The other office performed by the lime seems plainly to have been the setting up of conditions favorable to the growth in the soil of the micro-organisms by which the stores of organic nitrogen are gradually converted into available form through the process of nitrification. This is indicated by the fact that the giving of large quantities of available nitrogen in the fertilizers appears to have reduced the effect ascribable to lime, whereas this effect seems to have been augmented by fertilizers containing little or no nitrogen."

It may be said in general that lime as a soil amendment is more likely to be beneficial on heavy clay soil, in humid regions, where aëration is poor and the products of organic decomposition are very likely to be toxic to plants. In regions of low rainfall or sandy soils, lime is not so likely to be required as a soil amendment.

There are various chemical tests for determining the probable lime requirement of a soil, but the most reliable test is to apply lime experimentally and note results for at least two years.

References on fertilizers: —

Voorhees, E. B. (1898.) Fertilizers.

Lyon and Fippin. (1909.) Soils, pp. 267–386.

Bailey, L. H. (1911.) The Farm and Garden Rule Book, pp. 40–91.

Hopkins, C. G. (1910.) Soil Fertility and Permanent Agriculture.

Ohio Agr. Exp. Sta., Bul. 141; Circ. 104.

Maine Agr. Exp. Sta., Bul. 107.

Vt. Agr. Exp. Sta., Bul. 116; Circ. 7.

References on lime: —

Agriculture Lime. Conn. (Hatch) Agr. Exp. Sta., Bul. 163.

Lime and Liming. R. I. Agr. Exp. Sta., Bul. 46.

Chemical Methods of Ascertaining Lime Requirements of Soils. R. I. Agr. Exp. Sta., Bul. 62.

Liming Acid Soils. U. S. Dept. Agr., Farmers' Bul. 133.

Liming the Soil. Ohio Agr. Exp. Sta., Bul. 159.

Carriers of Lime. Ohio Agr. Exp. Sta., Circ. 123.

The Rational Use of Lime. Mass. Agr. Exp. Sta., Bul. 137.

Increasing the Yield of Corn. Tenn. Agr. Exp. Sta. Bul., Vol. XVII, No. 2, p. 46.

The Use of Lime upon Pennsylvania Soils. Penn. Dept. of Agr., Bul. 61. 1900.

CHAPTER XV

REGULATING THE WATER SUPPLY

A 50-BUSHEL corn crop requires 7 to 10 inches of water for the use of the plant, besides that to be allowed for run-off, seepage, and evaporation. In Nebraska, with a 29-inch rainfall, the division of this water between the four sources of losses is estimated as follows, when a 50-bushel crop is grown: —

Water required by the plants	8 inches
Water lost by run-off	3 inches
Water lost by seepage	2 inches
Balance lost by evaporation	16 inches
Total	29 inches

The proportion of total rainfall lost by the different means will vary with the region, but it is probable that in most cases evaporation is twice the amount required by the crop.

106. Not all evaporation is undesirable. Whenever the soil reaches its water-holding capacity, as is often the case in early spring, then it must be dried by evaporation before cultivation can be practiced. Run-off is desirable after the soil reaches saturation, if the run-off takes place in such a way as not to cause erosion, since the taking up of this water by the soil would increase the loss by drainage, and excessive drainage means a slow leaching of the soil. The amount of run-off necessary in order to care for excessive rainfall, or of evaporation necessary in order to dry out the soil, will vary with the rainfall. In fact, all the water above that actually used by the crop is excessive and must be disposed of in some way, as by drainage, run-off, or evaporation.

Even though the crop requires a relatively small proportion of the total rainfall, the crop often suffers due to the fact that this small proportion is required during a comparatively short period and in excess of the water-storing capacity of the soil.

Lyon and Fippin[1] give the following statement regarding the water-holding capacity of some soils: —

TABLE XXXIV

	Water Capacity		Amount of Available Water		
	Minimum Per Cent	Maximum Per Cent	Per Cent	Cu. in. per Cu. ft.	Inches per Acre, 4 ft.
Light sandy loam . .	3	8	5	122	3.4
Silt loam . .	15	25	10	218	6.0
Clay . . .	23	40[2]	17	274	7.6

Studies at the Nebraska station indicate the water requirements of a 50-bushel corn crop for the different months to be about as follows: —

TABLE XXXV

	Inches
January 1 to June 1	.00
June	.50
July	3.60
August	3.30
September	.60
October 1 to January 1	00
Total	8.00

[1] Lyon and Fippin. Soils, p. 158. [2] Assumed.

Most of this water is required during a period of five or six weeks, ranging from about July 10 to the end of August. On p. 65 it was pointed out that evaporation from the soil and loss from run-off probably equals or nearly equals the requirements of the plants in making a 50-bushel crop, or the total requirement by the crop, and evaporation from the soil, etc., for July and August probably amounts to 12 inches. This is twice the storage capacity of the soil and perhaps three times the amount usually available early in July. After the water stored in the soil is exhausted, if rains are delayed, the crop suffers, being greatly reduced, and this often happens even when abundant rains come later. The seasonal requirements of corn are illustrated by Fig. 24.

107. Three ways are open for regulating the water supply: (*a*) increasing the water-holding capacity of the soil; (*b*) conservation by preventing evaporation; (*c*) decreasing run-off during the growing season.

Since the water-storage capacity of soil is closely related to its physical composition, little can be done to improve this condition in a practical way. The addition of vegetable matter helps only to a limited extent.

A certain amount of evaporation can be prevented by cultivation, but how much has never been satisfactorily determined under field conditions. It is probable, however, that loss by evaporation of water that has reached a depth of 12 inches in the soil is very small, and that cultivation serves principally to prevent evaporation of moisture from rains that penetrate no deeper than 6 to 10 inches. Experimental results under field conditions to show the effect of cultivation give extraordinary variation. For example, at the Illinois Agricultural Experiment Station, plats of corn that were not cultivated but merely had the weeds shaved off gave as good results as an average of five years as when carefully cultivated, and similar results have been secured at other stations. (See p. 206.) On other occasions cultivation has apparently conserved sufficient moisture to improve the yield. The underlying principles have not been clearly worked out.

It seems apparent that a well-cultivated surface, with a good store of organic matter, will take up a moderate rain more readily and store a large percentage of it deep enough to protect from surface evaporation than will a hard and uncultivated surface; also, that when this moisture is stored continued cultivation will decrease the rate of loss from the upper 10 inches of surface.

EROSION

108. Effect of erosion. — Erosion affects the agricultural value of land in the two ways: first, by producing gullies and large ditches, thus increasing the expense of crop cultivation and resulting in the actual loss of some land; second, by reducing available fertility, through removing the surface.

In the latter case, the damage to productivity depends on the soil. In heavy clay soils, much of the available fertility seems to be in the surface 6 inches. On such soil productivity is often reduced for many years by turning up too much subsoil at one time with the plow. On the other hand, as pointed out by King,[1] in many regions, especially of low rainfall, the subsoil, even to several feet deep, is as productive as the surface soil. In a case of such surface, erosion would work little or no damage. However, in most of the regions where erosion is severe,

[1] King, F. H. The Soil, p. 29.

as in eastern United States, the soil is heavy in texture, the exposed subsoil not productive, and the loss of surface soil causes serious damage. When manure, mineral fertilizer, or lime is used, much of this added material remains in the plowed surface and erosion causes a direct loss of this material.

109. Causes of erosion. — In the corn-growing area of the United States — that is, from the Atlantic Coast westward to the 100th meridian — erosion is related to the amount of run-off water and to the condition of the soil at the time the run-off takes place. In the principal corn-growing States, north and west of the Ohio River, erosion is not serious. The land is generally level and rainfall not excessive. Also, during a part of the year the ground is frozen, and in June, July, and August, when about 40 per cent of the rainfall occurs, the land is in crop.

From Ohio eastward, however, the rainfall is heavier and cultivated land is more rolling, thus increasing the total run-off and erosion. From the western edge of the Corn Belt to the Atlantic Coast, erosion gradually increases. In Kansas and Nebraska, with level farming land, the rainfall is 25 to 30 inches and the total run-off about 3 inches. In the North Atlantic States rainfall is heavier, land more rolling, and the run-off is estimated at 40 to 50 per cent of the rainfall, which often amounts to a run-off of 20 inches or more. In the Southern States the most serious erosion takes place during the winter months. The soil is not frozen, is without a crop, and heavy rainfall occurs during this period.

The relation of cropping systems to erosion may be grouped as follows: —

(*a*) Land in grass erodes least.

(*b*) Land in stubble or small grain erodes more than (*a*).

(*c*) Land in cultivated crops erodes more than (*b*).

(*d*) Cultivated land not in crops erodes most.

110. Preventing erosion. — Since the character of the crop and the grade of the land both have a marked effect on the degree of erosion, they are the two principal means of preventing the same.

Land subject to severe erosion should be kept principally in grass crops and small grain, and never left longer than necessary without a growing crop. If a good supply of vegetable matter is maintained and deep plowing practiced, cultivated crops can often be grown on rolling land with little loss by erosion where otherwise the loss would be severe. It is often noted that new land just brought under cultivation does not erode, but as the humus supply decreases, erosion increases. Also, the plowing under of a heavy coat of barnyard manure or a green manure crop will often stop erosion where it is otherwise serious. Deep plowing enables the soil to take up water readily and give it up slowly, and in many cases deep plowing alone has been found to entirely prevent erosion.

The second method of preventing erosion is by decreasing the grade. This is usually done by terracing, causing the water to follow the contour of the hills at a low grade. The same effect is secured in some degree by plowing and planting with the contour of the hills.

To summarize: Erosion is better controlled when the land is in grass or small grain than when in a hoed crop. Sufficient organic matter and deep plowing decrease erosion. The land should not be left bare. The grade can often be decreased by terracing.

The most serious loss due to erosion is the constant removal of the accumulated organic matter of the surface soil.

DRAINAGE

111. Corn requires a thoroughly drained soil, both because it flourishes in a " warm " soil, and because it requires large amounts of available nitrates when making its rapid summer growth. On poorly drained land, even when such land is rich bottom soil, the corn plant will often have a yellow color indicating a need of nitrogen.

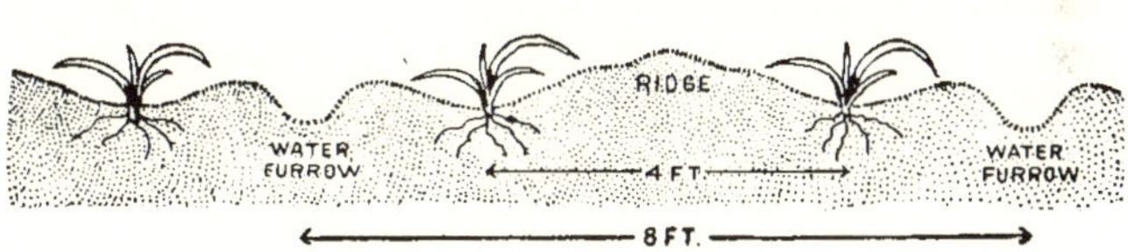

FIG. 40. — Plan of ridging land for surface drainage. Two rows on each ridge.

The water-logged soils interfere with bacterial activity and the normal nitrifying processes are prevented. Surface drainage for corn on very flat lands is often provided by plowing in narrow beds, 8 feet wide, and planting two rows of corn 4 feet apart on each bed.

Underdrainage is so thoroughly discussed in several soil texts that it is not necessary to take up the subject here.

SECTION IV

CULTURAL METHODS

CHAPTER XVI

PREPARATION AND PLANTING

So far in this book it has been the plan to discuss the fundamentals relating to the nature of the corn plant, its requirements, the conditions that must be met in the growing of corn, and methods of modifying the plant to improve yield or quality.

Having considered the above problems, the next step is to consider cultural methods. The basic principle in cultural methods is largely protection of the crop against unfavorable conditions that may arise, as draught, weeds, insect or parasitic enemies. The cultural method to be adopted in a particular case is the one that most effectually insures the crop, and at the least cost.

Cultural methods must vary with the local situation. In regions of high priced labor and level lands, extensive systems have been developed. In regions of low priced labor and small fields more intensive methods are practiced. The other crops to be grown, the character of the climate, the use of the crop, and many other factors all help to determine the most practical method to be adopted. As with other farm problems, the farmer himself must largely determine the cultural method to be used on his own farm.

THE OLD CORN STALKS

112. In the corn-belt and the Southern States, corn stalks are not harvested, but stand in the fields, to be plowed under the following spring. In the early days of corn culture in the middle west, the corn stalks were usually burned. The common custom was to break down the frozen stalks with a log or an iron rail and later when the ground had thawed, they were raked with horse rakes into long windrows, and burned. For a week or two in each spring, the sky would be lit up every night by the

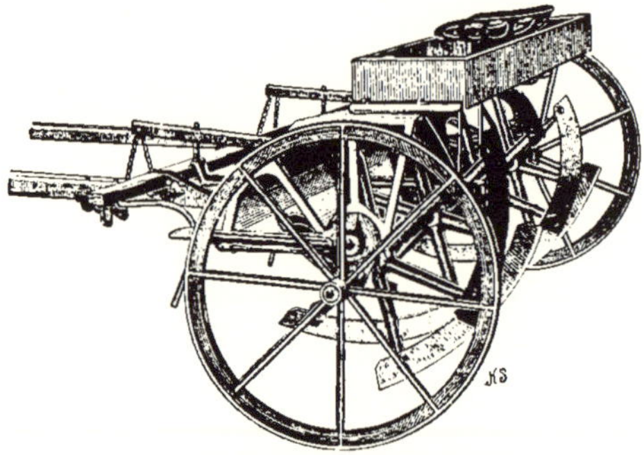

FIG. 41. — Two-row stalk-cutter.

great burning fields of corn stalks. This so rapidly reduced the organic matter in the soil that it soon became necessary to plow the stalks under, as is now the general custom, in order to obtain humus.

To prepare for plowing, the stalks are broken with a rail, as before, and then usually gone over with a sharp disk, to cut them up. The stalk-cutter is also in general use. This implement has heavy revolving cylinders set with knives that cut the stalks in twelve inch lengths. Where the stalks are heavy it is more satisfactory than the disk harrow, although the stalk-cutter is often followed also with a disk-harrow.

TIME OF PLOWING

113. When land is fall-plowed it is exposed more completely to the action of frost, thus giving a finer state of pulverization. This is often an advantage with heavy soils, but in light soils it may actually be a disadvantage. Also, when a cover crop is to be turned under, there is more time for decomposition when turned under in the fall.

When the soil is infested with the larvæ of injurious insects, fall plowing just as freezing weather begins will often destroy many of these. For early planted crops there is not always enough time for proper preparation of all the land in the spring, and it is good farm management to do part of the plowing in the fall.

Early spring plowing for corn, compared with late spring plowing, has not been the subject of extensive investigation. An experiment carried for a single season by Quiroga,[1] at the Ohio State University, showed an increase of about 7 per cent in the crop with early spring plowing overlate, and a marked increase in available nitrogen was found in the early plowed land throughout the season.

DEPTH OF PLOWING

114. From experiment stations some twenty-six tests have been reported on deep and shallow plowing for corn. These results cannot be regarded as very significant as a guide in specific cases, since the results were obtained under a great variety of conditions. They may be summarized as follows: —

Favorable to deep plowing	14
Favorable to shallow plowing	6
Indifferent results	6

[1] QUIROGA. Ohio State Univ. Bul., Series 8, No. 28.

There are no experiments to show the ultimate effect of following a system of continuous shallow plowing or continuous deep plowing, but practical experience has shown that land should be occasionally plowed deep (8 inches) to keep the surface in best mechanical conditions. Heavy soil requires deep plowing more often than do light soils. Probably a very heavy soil

FIG. 42. — Plowing under alfalfa sod in preparation for corn.

should be plowed deep once each year, while certain light soils, especially where rainfall is low, do very well with deep plowing every two or three years.

Hunt[1] summarizes certain experiments with deep and shallow plowing as shown on the following page.

It has been demonstrated many times, that if the soil has been kept in a good productive condition, that the preparation immediately before planting or even the system of cultivation after planting is not likely to have an important effect on the yield of the current crop. The crop secured does not depend so much on treatment of

[1] HUNT, THOS. F. Cereals in America, p. 220.

soil for the present crop, so much as the kind of treatment it has had for the last ten or twenty years. The kind of treatment to be recommended must consider more the future welfare of the land, than present benefits to be derived.

TABLE XXXVI

YIELD OF CORN IN BUSHELS

Station	Depth of Plowing Inches					
	2	4	6	8	10	12
Illinois	52.9	69.4	69.3	71.7		
Illinois	54.0		57.5		56.0	
Indiana (average 3 years)		39.5	40.5	42.3	41.8	42.0
Pennsylvania (average 3 years)		47.0	62.0	57.5	58.5	
New Hampshire[1]		14.2	26.2	29.4	28.2	
Alabama		24.1			24.2	
Minnesota		65.8	64.4	59.5[2]		
Ohio[3]		43.1		42.9		
Nebraska		38.5		31.0		

[1] Tons of green silage. Depths were 3, 5, 7, and 9 inches.
[2] Also subsoiled 6 inches deeper. [3] Depths 3 and 7 inches.

So far as tillage is concerned, as a factor in maintaining crop production, the following principles may be set forth:

That all land should occasionally be plowed 8 to 10 inches deep. On heavy land about once a year, but on lighter soil, and in rather dry regions, once in two or three years being sufficient.

The plowing should be done when the land is in proper condition to pulverize.

Quite thorough treatment with pulverizing tools, as

harrows, rollers, and cultivators, is essential to keeping the soil in good mechanical condition.

SUBSOILING

115. The subsoiler is a tool for loosening the subsoil without bringing it to the surface. While tools for this purpose have been in use for many years and have been generally tried out in all the principal agricultural regions, yet subsoiling is nowhere in general practice. General experience has confirmed results obtained at the Nebraska station, where, in a coöperative test with fifty-nine farmers for three years, beneficial results were obtained on soils having a heavy or impervious subsoil, but on loam subsoils the results were indifferent or injurious.

PREPARATION OF PLOWED LAND

116. The amount of fitting that must be given to land after plowing depends on the soil and the seasonal conditions. A good loam soil, plowed when in just the proper condition, may need very little fitting with the simplest tools, as harrow and float, in order to bring it to a proper mechanical condition. On the other hand, the same soil if plowed when too wet, or if when wet it had been tramped by stock in pasturing, would require more labor and a greater variety of tools for proper fitting. This emphasizes the importance of plowing only when the soil is thoroughly pulverized by the plow. Also, further pulverizing of the soil, with harrow or cultivator, is most easily accomplished within twenty-four hours or less after plowing, and one harrowing at this time may accomplish several times as much as a few days later, when the clods have dried.

There are certain heavy clay soils that always require a

great deal of fitting for good results. The best tool for pulverizing to a depth of several inches is the disk-harrow Where the land is stony or hard, a cutaway is more effective. On very stony or rough land, a spring-tooth is more practicable than the disk, or the ordinary cultivator can also be used to good advantage. For surface finishing, the spike-tooth harrow and weeder are used for pulverizing and the board drag and roller for further reduction.

Repacking the soil after deep plowing is an important function of all tillage in preparing the seed-bed. When the

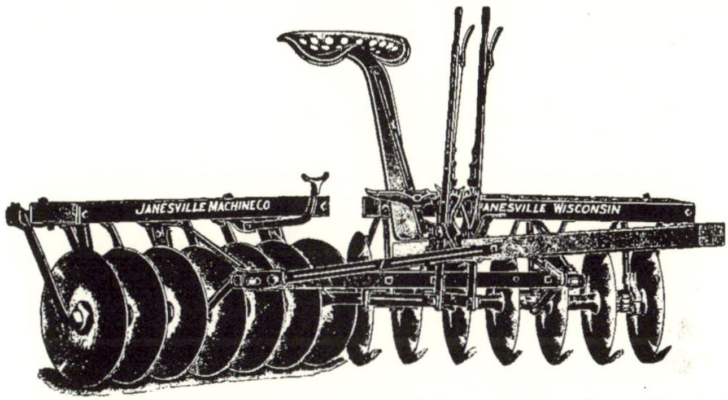

FIG. 43. — A modern disk-harrow. A tool that pulverizes the surface and packs the subsurface at one operation.

plowing is done long in advance, so that heavy rains may come, little attention need be given to repacking. A fairly compact seed-bed is desirable at planting time, though not so important with corn as with wheat.

A good method of repacking a loose seed-bed is to use either a subsurface packer, or quite as well a disk-harrow, set straight (no angle to disks) and loaded with sufficient weight to cut nearly through the furrow slice. These tools will pack the bottom of the furrow slice. To pack the surface, a roller or a smoothing harrow, or both, may be used.

Clearing of weeds is important in preparation. One principal advantage of early plowing is that more weeds may be germinated and destroyed before planting. While weeds are germinating rapidly, it is often an advantage to delay planting until the land can be entirely cleared, as it is much easier to destroy weeds before planting than afterward.

To sum up, it is important to plow the land when in just the right tilth for plowing, to pulverize thoroughly to repack when the seed-bed is loose, and to destroy weeds before planting.

PLANTING THE SEED

METHODS

117. (1) The seed may be "surface" planted, the land being prepared level and the seed planted in rows 1 to 3

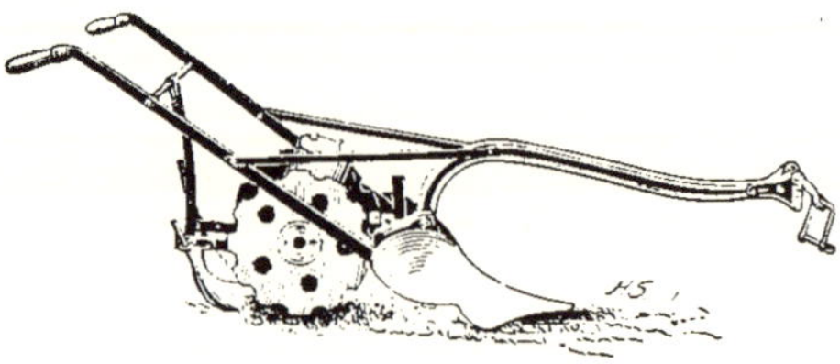

FIG. 44. — Combined lister plow and drill.

inches below the surface. (2) The planter may have a furrow opener, usually a pair of disks which open up a shallow furrow, the seed being planted in the bottom of this. (3) A lister may be used, which is essentially a double-moldboard plow throwing a furrow slice each way. The land is furrowed as deep as possible with the lister, the corn being planted in the bottom.

Surface planting is the method in common use on all heavy lands or in regions where rainfall is plentiful, being the common method in all the States east of the Missouri River. The "furrow opener," or disk planter, is also

FIG. 45. — A combined lister and drill. The land is not plowed in preparation for listing.

popular with many farmers, especially when it is desirable to drill, as in the growing of silage or fodder corn.

The lister came into vogue about twenty-five years ago, but it is used extensively only where the soils are rather light in texture (loam or sandy loam) and in regions of rather low rainfall. In central Nebraska, Kansas, and Oklahoma, one-half or more of the corn is listed. List-

FIG. 46. — Corn planter with disk furrow openers, to be used on plowed land.

ing is not practicable on land subject to washing, as the planting is likely to be destroyed by heavy rains. Also, in cold or wet soils the seed is likely to rot in the lister furrows, or growth of the young plants to be much retarded. Where listing is practicable, namely, in dry, warm soils, it is a very cheap method of producing corn, as the ground is not plowed before planting, though it is usually disked. Cultivation is simple and easy.

SOWING CORN FOR FORAGE

118. For coarse forage or soiling, corn is frequently sown broadcast or drilled thick with a grain drill. One to two bushels of seed are sown per acre. Usually a rather small early variety is used, rather than a tall or late variety.

FIG. 47. — Corn sown broadcast for forage. In above case was sown after wheat harvest.

Early sweet corn is well adapted for this purpose and is often sown in July after a small-grain crop has been harvested.

CHECKING AND DRILLING

119. When corn is to be surface planted it is usually "checked," that is, planted in hills and rowed both ways, thus permitting of cross cultivation. When corn is drilled on the surface, it is often difficult to keep weeds out of the

FIG. 48. — A two-row corn planter. Will drop in hills, rowing both ways, or in drills. Commonly called a check-rower.

row, as little soil can be thrown around the plants in cultivating. This difficulty is overcome in a large measure by using the furrow opener and placing the corn in a shallow furrow.

TIME OF PLANTING

120. Many experiments have been made on the time of planting, but the principal conclusion may be stated as finding an average range of about six weeks for corn planting. The very earliest and the very latest plantings are usually not so successful as those about midseason. For example, the Illinois station in 1890 made plantings from April 28 to June 9. The average yield of the corn

FIG. 49. — Special attachments for corn planter shoes.

planted in May was 73 bushels per acre, while the average yield of the three remaining plantings, one in April and two in June, was 63 bushels per acre.

Many experiments at other stations bear out the statement that there is a period of about three weeks for corn planting with equal chance of success, though there are occasional seasons when the very early or very late plantings are best. The optimum season is shorter in the North and longer in the South.

TABLE XXXVII

TIME OF PLANTING CORN IN CERTAIN REGIONS [1]

REGION	BEGINNING	GENERAL	ENDING	PLANTING PERIOD DAYS
Gulf States	March 15	April 5	May 10	55
Central States (Virginia to Kansas)	April 15	May 1	May 25	40
Northern States (New York to Minnesota) .	May 10	May 20	June 1	20

[1] U. S. Dept. Agr. Yearbook, 1910, p. 491.

The above table shows that the planting season begins about two months earlier in the Gulf States, as compared with the Northern States, but the total length of the planting season is about three times as long.

The average of the beginning of corn planting is also shown by the accompanying chart: —

Fig. 50. — Chart showing average date of planting corn in the United States.

The percentage of moisture in the crop at harvest time usually increases with the lateness of planting, after a certain date, as illustrated by the following data from the Illinois station:[1] —

[1] Ill. Agr. Exp. Sta., Bul. 20. 1892.

TABLE XXXVIII

DATA TAKEN AT HUSKING TIME

Date of Planting	Pounds to make One Bushel in Dry Corn	Percentage of Moisture in Shelled Corn	Bushels of Shelled Corn per Acre	Bushels of Air-dry Corn per Acre
April 25	69.9	14.0	52.6	50.8
May 2	70.8	14.6	52.6	50.4
May 9	70.9	14.8	50.7	48.5
May 16	74.4	17.0	53.3	49.7
May 23	80.0	19.3	57.9	34.1
May 30	96.8	24.0	40.0	37.5
June 8	97.9	23.9	43.9	37.5
June 13	127.8	31.5	25.2	19.4

DEPTH OF PLANTING

121. Corn is usually planted 1 to 4 inches deep. Results from several experiment stations are summarized as follows: —

TABLE XXXIX

PLANTING CORN AT DIFFERENT DEPTHS

Depth of Planting Inches	Yield per Acre in Bushels		
	Ohio[1] Average 6 Years	Indiana[2] Average 6 Years	Illinois[3] Average 5 Years
1	56.6	38.6	78.0
2	51.2	39.2	72.0
3	46.8	37.8	65.0
4		28.8	69.0
5			61.0
6			60.0

[1] Ohio Agr. Exp. Sta., Rpt. *1890:* 87.

[2] Ind. Agr. Exp. Sta., Bul. *64:* 5. 1897.

[3] Ill. Agr. Exp. Sta., Bul. *31:* 353.

In no case has the average yield been increased by planting more than 2 inches deep. In heavy soils, such as of the Ohio station, shallow planting was decidedly better, while in loose loam soil, at the Illinois station, the depth of planting did not vary results so much. Also, when the soil is warm and dry the corn should be planted deeper than when the soil is cold. In two years out of seven at the Ohio station, when the soil was drier than common, the 3-inch plantings gave the best results.

Some persons have thought that deep planting would establish the roots deeper in the soil. It has been found, however, that the roots spread out at about the same depth, no matter what the depth of planting. Ordinarily the roots spread out about 1 inch below the surface.

It would seem, then, there is no object in planting corn deeper than is necessary to insure plenty of moisture for good germination.

RATE OF PLANTING

122. The customary rate of planting varies with soils and climate. In the South the corn rows are often 5 feet apart and the hills 4 feet apart, with two stalks to a hill. The rate of planting increases toward the North. Customary rates are as follows: —

TABLE XL

Region	Distance Apart of Hills	Plants per Hill	Plants per Acre
Gulf States	4′ + 5′	2	4,000
Middle States (Virginia to Kansas)	3′8″ + 3′8″	2–3	9,000
Northern States (New York to Minnesota)	3′6″ + 3′6″	3–4	12,000

The rate of planting is partly regulated by the size of plant. Plants in the Gulf States are about twice as large as in the Northern States, due in part to climate and also to the longer growing season.

It has been shown, however, that for a given place the rate of seeding within wide limits does not have a marked effect on yield. An experiment regarding this point was conducted by the Illinois Agricultural Experiment Station.[1]

Fig. 51. — A Southern method of planting on poor soils. Rows wide apart, and a crop of peanuts between. For soil improvement cowpeas are sometimes grown between.

For three years corn was planted at rates varying from 5,940 to 47,520 kernels per acre. The maximum yield was obtained with 11,573 kernels per acre, though almost as good yields resulted when 15,840 or 23,760 kernels were

[1] Ill. Agr. Exp. Sta., Bul. *13* : 410.

planted. The average yields were 81, 77, and 76 bushels per acre, respectively.

At the Nebraska station, corn was planted in hills 3 feet 8 inches apart each way, the stand varying from one to five plants per hill.

TABLE XLI

AVERAGE RESULTS FROM PLANTING CORN AT VARIOUS RATES FOR SIX YEARS (1903–1908), NEBRASKA STATION[1]

PLANTS PER HILL	YIELD PER ACRE BUSHELS	AVERAGE WEIGHT OF EAR OUNCES	NUMBER OF EARS PER 100 PLANTS	NUMBER OF TILLERS PER 100 PLANTS	TWO-EARED PLANTS PER 100	BARREN PLANTS PER 100
1	48.3	10.5	161	138	13.3[2]	3.0
2	67.7	10.6	115	60	4.9	4.8
3	75.5	9.4	95	25	2.4	6.9
4	76.7	8.2	82	10	.8	8.3
5	76.3	7.4	77	3	1.1	10.8

[1] Nebr. Agr. Exp. Sta., Bul. *112* : 30. 1909.
[2] Four years only.

There was practically no difference in yield when three, four, or five plants were grown to the hill.

ADJUSTMENT OF CORN PLANTS

123. As the number of plants increased, the size of ear and the number of ears decreased, while the number of barren plants increased. One stalk per hill produced 64 per cent and two stalks per hill 90 per cent as much grain as did three stalks per hill, due principally to the increased size of ear and number of tillers producing ears and to the decrease in number of barren plants. It is evident that the corn plant is capable of a wide range of adjustment

ECONOMIC VALUE OF TILLERS

124. The question often arises as to whether tillers should be pulled when they appear in abundance. Data were taken at the Nebraska station for five years, and in every case the yield was decreased by removing tillers. For three years the corn was planted at different rates, the data being summarized as follows: —

TABLE XLII

EFFECT ON YIELD OF GRAIN OF REMOVING TILLERS FROM CORN
THREE-YEAR AVERAGE (1906–1908)

Number of Plants per Hill	Average Yield in Bushels per Acre			Average Yield of Stover in Pounds per Acre			
	Tillers on	Tillers Removed	Difference in Favor of Tillers	Tillers on	Tillers Removed	Increase Due to Tillers	Percentage Decrease Due to Removing Tillers
1	45.9	31.8	14.1	5,061	2,208	2,853	56.3
2	66.1	56.4	9.7	5,127	4,200	927	18.1
3	69.6	64.4	5.2	5,115	4,687	428	8.3
4	73.2	71.4	1.8	5,801	5,602	199	3.4
5	76.7	72.6	4.1	6,043	5,987	56	.9

Tillers appear to develop in response to the needs of the crop, in an attempt to bring the stand up to normal. When the stand is maximum, few tillers develop. The occasions are certainly very rare when it would pay to remove tillers.

OTHER FACTORS AFFECTING PRODUCTION OF TILLERS

125. On some soils tillers do not develop even when the planting is thin. When early growth is slow or retarded, as on heavy or cold clay soils, there is not sufficient stimulus early in the life of the plant to start the tillers.

RATE OF PLANTING ON DIFFERENT SOILS

126. **On good soils** it is generally recognized that planting should be thicker than on poor soils. This is shown by data obtained by the Illinois station.[1] In a series of tests on different soils, corn was planted in hills at various distances apart and two or three stalks per hill. Grouping the data so as to include all fields yielding more than 50 bushels per acre in one class, and all yielding less than 50 bushels in the other class, the following results are obtained:

TABLE XLIII

RATE OF PLANTING AND YIELD ON SOILS PRODUCING MORE OR LESS THAN 50 BUSHELS PER ACRE

Region	More than 50 Bushels per Acre		Less than 50 Bushels per Acre	
	Two Kernels per Hill	Three Kernels per Hill	Two Kernels per Hill	Three Kernels per Hill
Northern Illinois . . .	57.9	68.5	41.4	42.4
Central Illinois . . .	59.8	62.8	43.2	40.9
Average	58.8	65.6	42.3	41.6

On productive soil the yield was increased by the thicker planting; but on the poorer soil two kernels per hill evidently furnished the maximum stand, as no further increase was secured by three kernels per hill. Data from the Indiana station show that in dry seasons the

[1] Ill. Agr. Exp. Sta., Bul. *126:* 366–377. 1908.

thin plantings give the best results, while in favorable seasons the reverse is true:[1] —

TABLE XLIV

EFFECT OF SEASON ON YIELD AND PERCENTAGE OF GRAIN

Stalks Inches Apart	Seasonable, 1888–1891		Ears Percentage	Dry, 1893–1894		Ears Percentage
	Bushels Corn	Pounds Stalks		Bushels Corn	Pounds Stalks	
19½	49.76	3,617	49.1	22.07	3,092	33.3
16	54.05	4,065	48.2	21.27	3,143	32.2
14	57.79	4,158	49.3	19.39	3,762	26.5
15	57.81	4,201	49.6	14.28	5,204	16.1
11	59.14	4,960	45.5	13.80	4,360	18.1

This also indicates that in semiarid regions, as central Nebraska or Kansas, the regular practice should be rather thin planting.

METHOD OF DISTRIBUTION OF PLANTS

127. At the Illinois station, hill planting was compared with drill planting at various rates per acre. For example, four plants would be planted in hills every 48 inches, in comparison with two plants every 24 inches or 1 plant every 12 inches. The conclusion was that it made no difference in what manner the seed was distributed, so that approximately the same number of plants per acre were grown in each case.

At the Nebraska station, a uniform distribution of three grains per hill was compared with distributing the

[1] Ind. Agr. Exp. Sta., Bul. *64* : 4.

seed in different amounts per hill but planting the same number per acre. The uniform distribution had a slight advantage, but not enough to indicate that the ordinary variation in dropping in corn planters would materially affect the yield.[1]

WIDTH OF ROWS

128. Width of rows is an important consideration, since the amount of labor required in planting and cultivating an acre is directly related therewith.

TABLE XLV

Distance Apart of Rows Feet	Rods of Travel in Cultivating One Acre	Percentage Increase in Labor
4.0	650	
3.5	754	16 per cent increase over 4 feet
3.0	880	17 per cent increase over 3.5 feet 35 per cent increase over 4 feet

Numerous experiments have not shown a practical advantage in having rows closer than 36 inches in the northern limit of corn-growing States, 42 inches in the central corn States, and 48 inches in the Southern States, when the standard type of corn for the region is grown primarily for grain. A small early variety may be planted closer.

When the corn is grown primarily for silage or fodder, somewhat closer planting will give a greater yield of forage.

[1] Nebr. Agr. Exp. Sta., Bul. *112* : 35.

YIELD OF FORAGE

129. When yield of forage is considered, numerous experiments have shown that the yield of forage increases with the rate of planting up to a point about twice that required for maximum yield of grain. The following data illustrate: —

TABLE XLVI

YIELD OF GRAIN AND STOVER WHEN CORN WAS PLANTED AT DIFFERENT RATES. THREE-YEAR AVERAGE. ROWS 3 FEET 8 INCHES APART. ILLINOIS STATION[1]

Rate of Planting Kernels per Acre	Bushels of Shelled Corn per Acre	Tons of Stover per Acre	Ratio of Shelled Corn to Stover
5,940	55	2.5	100 : 16
9,504	72	2.9	100 : 140
11,880	81	3.0	100 : 130
15,840	77	3.1	100 : 140
23,760	76	3.7	100 : 174
47,520	59	4.8	100 : 290

[1] Ill. Agr. Exp. Sta., Bul. *13* : 410.

EFFECT ON COMPOSITION

130. The principal effect on composition when the rate of planting is increased is the change in ratio between percentage of ear and stalk. By referring to Table XLVI, last column, it will be seen that the proportion of stalk to ear increases as the rate of planting increases, there being more than twice the proportion of stover with the thickest planting as compared with the minimum ratio (11,880 kernels). The comparative analysis of stover and grain as summarized by Jenkins and Winton is given in the following table: —

TABLE XLVII

COMPOSITION OF STOVER AND GRAIN IN CORN. WATER-FREE BASIS

	NUMBER OF ANALYSIS	ASH PER-CENTAGE	PROTEIN PER-CENTAGE	FIBER PER-CENTAGE	NITRO-GEN-FREE EXTRACT PER-CENTAGE	FAT PER-CENTAGE
Fodder . . .	35	4.7	7.8	24.7	60.1	2.8
Leaves . . .	17	7.9	8.6	30.6	51.0	1.9
Husks . . .	16	3.5	5.0	32.2	57.9	1.4
Stalks . . .	15	3.6	5.9	34.8	64.1	1.6
Stover . . .	60	5.7	6.4	33.0	53.2	1.7
Grain . . .	208	1.7	11.7	2.4	78.1	6.1

In well-developed corn planted at proper distance for maximum yield, the weight of shelled corn will be almost equal to the weight of stalk. Increasing the rate of planting has very little effect on the composition of either grain or stalk, but, as the proportion of stalk to grain increases, it is evident that the analysis of the whole plant will show a decreased percentage of protein and fat and an increased percentage of fiber. The total protein per acre, however, will increase. Silage from very thickly planted corn will not be so rich in percentage of protein and fat, but the total yield per acre will be greater.

By reference to Table XLIV it will be seen that the rate of planting has more effect on percentage of ears in a dry season than in a seasonable year. The same would be true on poor soil.

CHOICE OF A VARIETY

131. There are probably one thousand named varieties of corn. This very large number of varieties, many of

[1] U. S. Dept. Agr., Office Exp. Sta., Bul. 77. 1892.

which are of only local importance, makes rather confusing a study of experiments, in order to select the best varieties. In some cases a number of varieties have had a common origin and for a general discussion might be grouped to-

Fig. 52. — Rough division of the United States into corn regions, according to the types of corn grown.

gether. There are other groups, originating from widely different sources, which are yet very similar for all practical purposes.

The eastern half of the United States, where most of the corn is grown, may be roughly divided into large regions, within which certain types and varieties predominate to a greater or less degree.

Elevation must always be considered in selecting a type. For example, the coast plains of North Carolina would probably require a type similar to that suitable to the Gulf States, while the mountain regions would require a type

FIG. 53. — Prolific varieties of corn produce from two to six ears per stalk. They are adapted principally to the cotton belt. (Cockes' prolific.)

normally adapted to a region as far north as Ohio. Thus, in North Carolina, above 2800 feet, flint varieties are recommended — the type of corn most common in the New England States. Other local considerations enter in, but in general the following varieties have been found satisfactory in the regions indicated: —

Natural divisions

Section No. 1. Gulf States. Prolific varieties bearing 160 to 200 ears to 100 stalks, on the average, give better results than those bearing only single ears. Among the best of these are: —

Mosby	Sanders	Albemarle
Cocke's Prolific	Blount	Marlboro

Large-eared varieties are: —

St. Charles White	Boone County White

Section No. 2. In this region large single-ear varieties share about equal importance with prolific varieties. In addition to the prolific varieties named above, we find such varieties succeeding as: —

For good fertile land: —

Boone County White	St. Charles White
Huffman	White Pearl
Leaming	Hickory King

For poorer soils and upland: —

Hickory King	Sanders
Leaming	St. Charles White (Early Strains)

For high elevations: —

Eight-row Flint

This region partakes about half and half of the varieties common to the regions north and south of it.

Section No. 3. This is the "Corn Belt." Only large single-ear dent varieties are grown. South of this belt the dent corn is mostly white in color, but in the Corn Belt yellow corn is as popular as white. The leading varieties are:—

Yellow	*White*	*Early varieties*
Leaming	Silver Mine	Pride of the North
Ried's Yellow Dent	Boone County White	Early Calico
Riley's Favorite	Johnson County White	White Cap
Legal Tender	St. Charles White	

Leaming is probably the most extensively cultivated corn in the United States, being not only a universal favorite as a field corn, but also grown extensively for silage corn. Silver Mine is probably second in importance.

Section No. 4. This is more of the nature of a small-grain region, but corn culture is increasing. A few years ago flint corns predominated, but in recent years early dent corns have been developed and have largely replaced the flints.

Dent varieties	*Flint varieties*
Pride of the North	King Philip
Minnesota No. 13	Smut Nose
Wisconsin No. 7	Eight-row Yellow
Early Huron	Hall's Gold Nugget
White Cap	

Section No. 5. Flint corns are grown principally, though on the best soils below 1000 feet elevation. The early dent varieties share about equal popularity with the flints. Above 1000 feet elevation, flints are almost universal.

Flint varieties	*Dent varieties*
Eight-row and Twelve-row Yellow Flint	Pride of the North
King Philip	White Cap
Canada Smut Nose	Hall's Gold Nugget
	Various acclimated local varieties

In this section, one-third to one-half of the corn is grown for silage. For this purpose the seed is usually

Fig. 54. — Four ears in center are Sanford white flint, the longest type of cultivated corn. On right and left are shown typical ears of dent and flint, for comparison.

purchased and large varieties are used that do not ripen grain but are barely mature enough for silage when frost comes. Leaming is the favorite with Hickory King, Eureka Ensilage, Burrill and Whitman, and Evergreen Sweet following. In fact, almost all the large dent varieties are used to some extent for ensilage on the lower elevations, while flints are grown on the higher lands.

The importance of using acclimated seed has already been pointed out (page 117). Acclimated native seed should always be used for grain growing; and even for ensilage, while it is not necessary that the grain should mature, a better quality of silage is secured if the climatic change is not too great.

PREPARING SEED FOR PLANTING

132. In the more humid part of the Corn Belt, corn is very likely to decrease in germination. This necessitates some precautions in curing the seed corn. In regions where the fall and winter climate is clear and comparatively dry, there is less difficulty, but abnormal conditions occur often enough to justify special care of the seed corn as a regular practice.

CAUSES OF POOR GERMINATION

133. Slow or imperfect drying of the mature corn, often accompanied with freezing, seems to be the principal cause of deterioration of vitality in the germ. When corn is first " ripe " the kernels will usually contain about 30 per cent moisture. This would be about September 15 to October 1 in the Northern Central States. If the weather is dry and favorable, the grain should dry down to about 20 per cent moisture in the course of four to six weeks. If the climate is fairly dry, the corn should then remain in a good germinating condition either on the stalks or in good dry storage.

The principal cause of loss in vitality seems to be failure to dry out properly upon becoming ripe. It is not necessary for the corn to be frozen to lose vitality, as it deteriorates at ordinary temperatures during the three months following maturity if not fairly dry. If freezing occurs, the loss is increased. A freezing temperature occurring when the grain still contains a high percentage of moisture may practically destroy vitality.

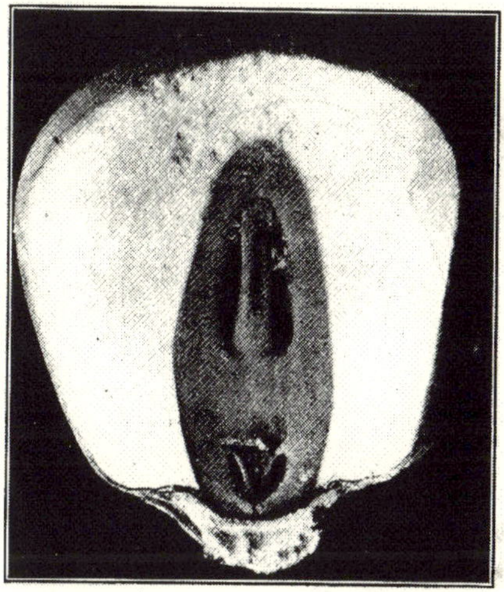

Fig. 55. — Corn kernel split to show germ, which is the dark-colored body within the white, and extending nearly the length of the kernel. The main outer part of the germ is the *Scutellum*, secretes an enzyme that reduces the starch for use of young plant. The column-like body in the upper half is the *Plumula*, develops into young plant. The body at the lowest point is the *Radicle*, or root of young plant.

Any cause that delays the proper drying of the corn after maturity will result in poor seed corn. In many cases, growers are using varieties too late in maturing or not well acclimated. Deep-kerneled types are more likely to lose in vitality than shallow-kerneled corn. Varieties with large, sappy cobs are always slow in drying.

STORING SEED CORN

To insure good seed corn, the ears should be collected as soon as mature and dried. Methods of drying are discussed elsewhere.

GERMINATION TESTS

134. If seed corn has been properly saved, there will be no occasion for making germination tests. It is much cheaper to save the seed properly than to make germination tests. Whenever seed is to be selected from a supply, the quality of which is doubtful, careful germination tests should be made.

The general test

135. A general test should be made first. Choose 100 ears at random and remove three kernels from each at different parts of the ear, as butt, tip, and middle.

A good germinater is made by using two pie tins or dinner plates. Fill one with sand, sawdust, or soil. Place

FIG. 56. — A simple germinater for testing seed corn. The corn is placed between damp cloths or blotters.

a cloth on this and spread out the kernels to be germinated. Place a second cloth over the seeds and wet down. Then invert the second pie tin or dinner plate over the first so as to make a moist chamber within. Keep moist and in a warm place. Six days is sufficient time to allow for germination. If 90 per cent or more of the seeds show good strong sprouts, it is doubtful if it would pay to make a germination test of each ear separately.

The ear test

136. When the preliminary test shows germination to be low or a high percentage weak, it will pay to germinate each ear separately.

There are several " seed testers " on the market adapted for this work, but satisfactory germinaters can be made

FIG. 57. — Making a germination test. The rack contains 100 ears, corresponding in number to the squares in the germination box.

at home. Usually a series of shallow trays are made and filled with sawdust or sand. A cloth is laid on top marked off in two-inch squares, and each square is numbered. Twenty inches square is a convenient size, though some prefer a tray twice to five times as large. The ears to be tested are laid out on shelves in sets of ten. The ears are then taken in order, six grains removed, and these grains placed in the corresponding square on the cloth.

It is well to take two kernels from the butt, two from the middle, and two from the tip, of the ear. When a tray has been filled, the grains are covered with a second cloth and a little sawdust on top and thoroughly wet down. When all trays are filled they are stacked up in a warm place and wet once a day for five or six days. All ears that have not shown a strong germination by this time should be discarded.

IMPORTANCE OF STRONG VITALITY

137. It should be emphasized that only ears showing a strong, quick-growing germ should be used. C. P. Hartley records a typical experiment illustrating this point.[1]

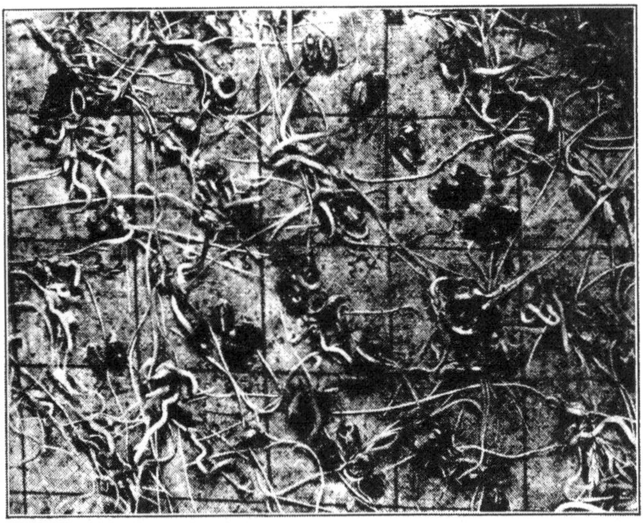

FIG. 58. — Difference in germination of ears. In each square are six kernels, each from a different ear.

[1] HARTLEY, C. P. The Seed Corn Situation. U. S. Dept. Agr., Bur. Plant Indus., Circ. No. 95. 1912.

In November two bushels of seed corn were selected, one bushel being placed in a corn crib and the other in a dry seed room. Germination was about equally good in both cases, but the plants from the seed kept in the dry house were stronger and the yield averaged five bushels more per acre.

GRADING SEED

138. The corn planter cannot be adjusted to uniform dropping of seed unless the kernels are uniform in size.

FIG. 59. — Three rows on left from single ear of good seed corn. Three rows on right from single ear of poor seed corn.

Some growers sort the seed ears into two or three lots, according to size of kernel. In some cases "sorters" are used, consisting essentially of a pair of screens that take out both the extra large and the extra small kernels.

CALIBRATING THE PLANTER

139. The dropping devices on planters are of three types, known respectively as (1) round hole drop, (2) round hole accumulative, and (3) edge drop accumulative. The first type represents the earliest type of dropper plate, when it was attempted to regulate the number of kernels per hill by the size of hole in the dropper plate; the hole being large enough to take two, three, or four grains at a time. In both the accumulative drop forms, the hole

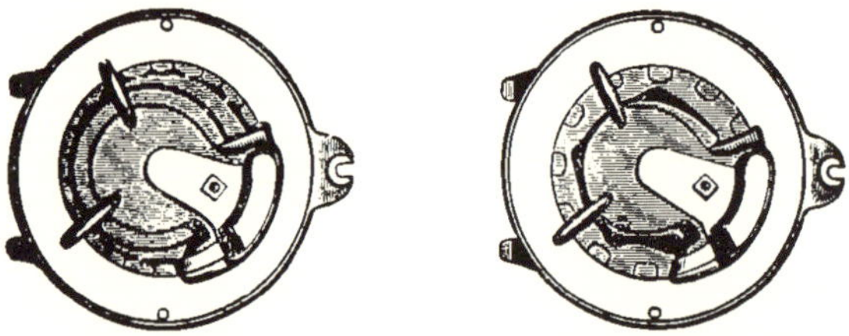

EDGE DROP FLAT DROP

FIG. 60. — Two types of planter plates for dent corn. The edge drop is considered best where the corn is sorted to uniform size, and flat drop where the seed is not uniform.

is large enough to take only one kernel at a time, the desired number of kernels being accumulated one at a time in a pocket and then dropped. The latter method is considered more nearly accurate when the seed has been well sorted. Before starting to plant, care sould be taken to see that the dropper-plate holes are of the right size for the seed used.

CHAPTER XVII

THE PRINCIPLES OF INTERCULTURE

TILLAGE MACHINERY

A GREAT variety of tools has been developed especially adapted for the tillage of corn. For the first cultivation of drilled or checked corn, the common smoothing harrow is often used. It is an excellent tool for this purpose as

FIG. 61. — The weeder. A very useful tool on loose soil, for cultivating corn the first four weeks. Cultivates three rows at a time.

it works a wide swath and kills young weeds effectively. One disadvantage is that it carries considerable trash especially where there are many large corn stubbs in the land. In this case the weeder is much better than the spike tooth harrow, as it clears of trash and does less injury to the young plants. When the weather is dry and the plants tough, a weeder may be used until the corn has reached the height of twelve inches.

FIG. 62. — The simplest type of one-row cultivator, in extensive use throughout the corn belt.

The corn cultivator has undergone a rapid evolution in the past fifty years. The first horse cultivators were single

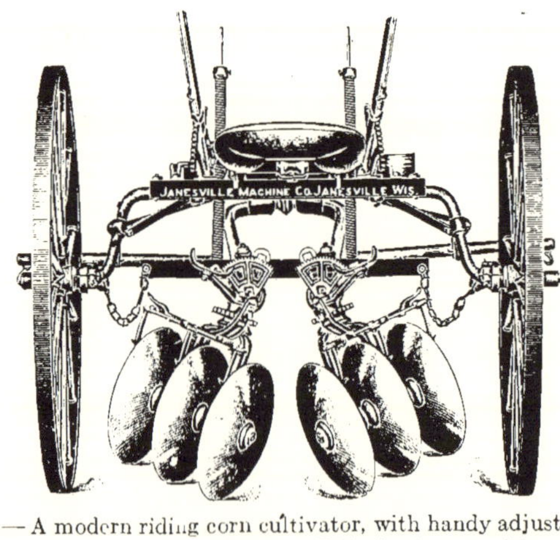

FIG. 63. — A modern riding corn cultivator, with handy adjustments and attachments, to readily adapt for all kinds of corn cultivation. Disk gangs attached.

shovel plows, consisting of a very broad mold-board shovel mounted on a beam, with handles to guide. Later two narrower shovels were substituted for the single broad shovel. Though this was an improvement, it was still necessary to go twice in each row for thorough cultivation.

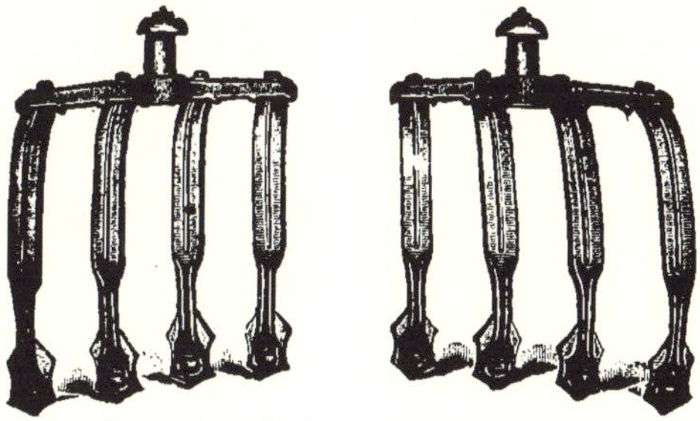

Fig. 64. — Spring-tooth attachment.

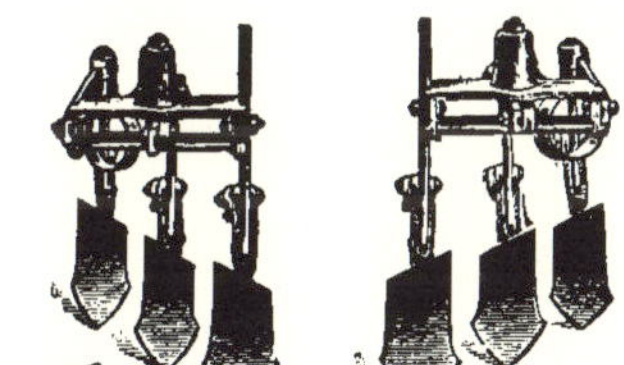

Fig. 65. — Shovel attachment.

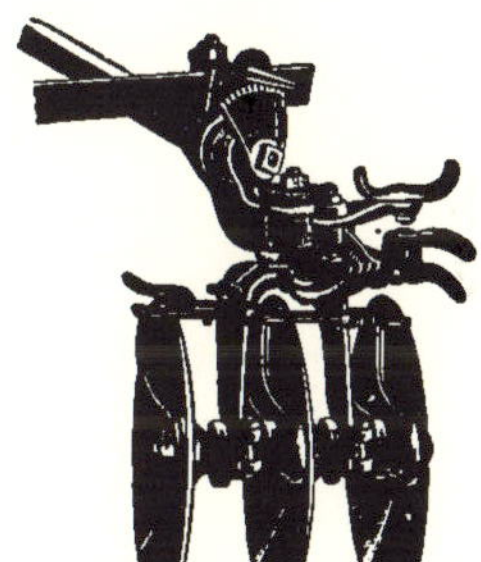

Fig. 66. — Cut showing angle and tilt adjustments.

Later two of these double shovel plows were rigged on a two wheel sulky, thus enabling the operator with two horses to cultivate both sides of a row at one time. The corn cultivator is still built essentially on this principle with many types of shovels and improvements for ease in controlling as illustrated in Figs. 63–66.

Modern cultivators may be fitted with four to eight shovels, the size of the shovels decreasing as the number increases. The six or eight shovel type is usually preferred where the ground is in good tilth and the weeds small.

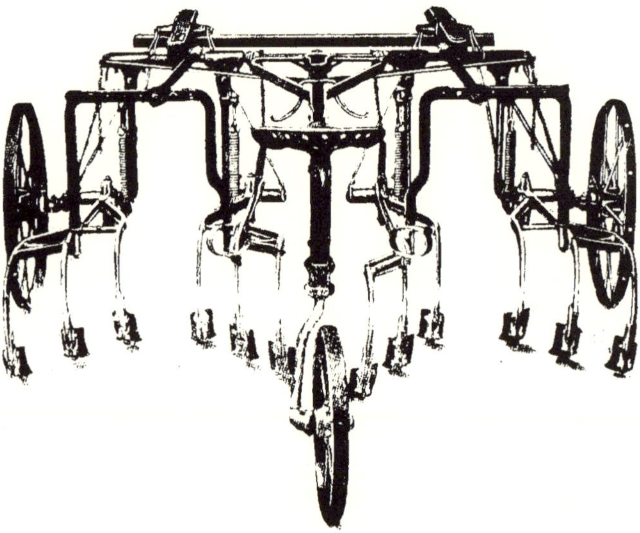

FIG. 67. — Two-row corn cultivator for three horses.

Where the ground is hard and the weeds large, so that the land must be plowed rather than cultivated, the large four shoveled type is more effective. On stony land the spring tooth gang is often preferred. Also most standard riding cultivators may be fitted with disk gangs. Disk cultivators do excellent work in the hands of a skilled operator. They are especially desirable when

Fig. 68. — Two-row corn cultivator for two horses, in operation.

the soil is in poor physical condition and needs pulverizing.

Two-rowed cultivators adapted for use with either two or three horses are now in general use. If two-row cultivators are to be used, the rows should be straight and uniformly equal distances apart. With the two-row cultivator it is not possible to do as careful work close to

FIG. 69. — Late cultivation of corn, with narrow tooth plow.

the row as when a single row is worked at a time. On the other hand, when the corn is clean in the row it may do all that is necessary in half the time.

One horse cultivators are not used much in corn cultivation, except occasionally for late cultivation where the plants are too high to straddle.

For listed corn a variety of tools has been specially devised. A spike tooth harrow is often used to level the ridges slightly when the corn first comes up. Then a tool such as illustrated in Fig. 70 is sometimes used or, more commonly, a two-row tool of the type illustrated in Fig 71. The first time over, the disk followers are usually set to throw out, as shown on the right of the figure, with a shield to protect the young corn and a pair of small

FIG. 70. — Tool for cultivating listed corn the first time over.

shovels to work in the bottom of the furrow. Later the disks may be set wider apart and set to throw toward

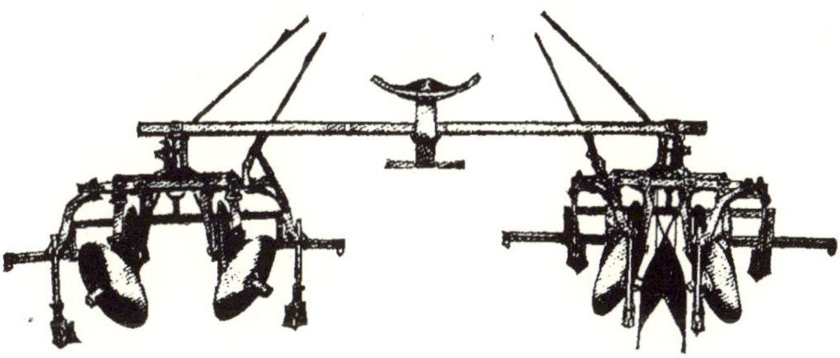

FIG. 71. — Two-row listed corn cultivator.

the corn. The shovels may be adjusted to suit conditions.

FIG. 72. — Two-row listed corn cultivator in use. The plants are just up. Listed corn is cultivated with this tool two or three times.

140. Jethro Tull said, "Tillage is manure," and this axiom has been more or less accepted and inculcated into our theories regarding the interculture of hoed crops. In the case of small grain crops, which are sown thickly enough to fully occupy the land, benefit has rarely been derived from interculture. With crops which are planted wide apart and which never fully occupy the intervening ground, it has been found profitable to give sufficient interculture to prevent the growth of weeds.

How much more interculture may benefit the crop than by keeping down weeds is a debated question. Various reasons have been advanced to account for the benefits of interculture and these may be summarized as follows:

To destroy weeds.

To conserve moisture.

To reduce run-off of rainfall by keeping the surface loose and porous.

To aërate the soil.

To increase availability of plant food.

The relative importance of each of the above functions of interculture will vary according to locality and season. Interculture to aërate the soil and to free fertility may be important on certain heavy clay soils in a humid region, but negligible on more porous soils or in a dry region. Where torrential rains occur during the growing season, it is important to have the surface in a porous, granular condition.

In general, however, the conservation of moisture and the destruction of weeds are properly advanced as the principal objects of interculture. Of all objects, the destruction of weeds appears to be paramount. This conclusion is arrived at as the result of numerous experiments, which have shown that keeping down weeds by shaving off has given almost as good results as when the soil was given good cultivation.

METHODS OF TILLAGE COMPARED

141. In the following tables are shown the results of three of the above-mentioned experiments under very different climatic and soil conditions, namely, New Hampshire, Illinois, and Utah. All give the same general conclusion — that culture beyond the destruction of weeds has not given much increased yield.

TABLE XLVIII

Results at Three Stations with Different Methods of Cultivating Corn. New Hampshire Station (Bul. 71, 1900)

Kind of Culture	Yield Bushels per Acre
No culture, weeds permitted to grow . .	17.1
Shallow, 14 times	80.6
Shallow, 5 times (ordinary culture) . .	79.1
Deep, 5 times	69.7
Mulch, covered with swamp hay . . .	56.1

Illinois Station (Bul. 31, 1894)

Kind of Culture	Average Yield for Five Years Bushels per Acre
None, weeds scraped with a hoe . . .	68.3
Shallow, about four cultivations . . .	70.3
Deep, about four cultivations	66.7
Shallow, about eight cultivations . .	72.8
Deep, about eight cultivations	64.5
None, weeds allowed to grow	

Utah Station (Bul. 66)

Kind of Culture	Average Yield for Eight Years Bushels per Acre
None, weeds pulled by hand	51.8
Scuffle hoe (scarified)	58.8
Shallow tillage, 1¼ inch	52.9
Medium tillage, 2¼ inches	57.3
Deep tillage, 3¼ inches	57.4
Mulched with soil	55.8

It has been shown by numerous experiments on bare soils that a mulch of straw or of dry loose earth would conserve considerable moisture. It has also been pointed out heretofore (page 67) that the need of water is the most common limiting factor in corn production. Reasoning from this, it seems that interculture should play an important part in conserving moisture and this increasing yield, but practical experiments fail to show such increases.

WATER-LOSS FROM FALLOW SOIL

142. For three months (April, May, and June) the prospective cornfield is essentially a bare field, exposed to wind and sunshine; and it is to be expected that early plowing and maintenance of a soil mulch will conserve moisture during this period.

At the Wisconsin station adjacent plots of land were plowed in early spring seven days apart. During this interval of seven days the unplowed plot lost 1.75 inch of water, while the plowed plot had actually gained moisture in the first 4 feet, probably due to capillary water from below.

Widstoe[1] states that "Fortier, working under California conditions, determined that cultivation reduced the evaporation from the soil surface over 55 per cent." At the Utah station similar experiments have shown that saving of soil moisture by cultivation was 63 per cent for a clay, 34 per cent for a coarse sand, and 13 per cent for a clay loam.

EVAPORATION UNDER CORN CROP

143. When the corn becomes large enough to shade the ground, which will be soon after the time that interculture begins, most of the conditions causing loss of soil moisture in fallow soils will have become to a large degree ineffective. Wind, the most potent cause of soil drying, is almost nil at the soil surface; direct sunshine is cut off, the soil being in shade part of the time; and humidity is higher. At the Nebraska station, jars of water set in wheat fields level with the soil surface lost practically no water.

Another important factor in preventing loss of soil water by evaporation is the spread of roots near the surface. (See page 27.) If there is no rain, practically all water moving upward from the subsoil is intercepted by these roots and used by the plants. If there is rain, the surface moisture is soon reduced by the surface roots to a point where upward capillary movement is retarded.

From the above, it appears that interculture of the corn crop can do very little toward conserving moisture.

THE EFFECT OF WEEDS

144. A crop of weeds will not only take out moisture, but also consume available plant food. As plant food in

[1] Widstoe, John A. Dry Farming, p. 155.

available form is usually more limited than the water supply, the consumption of plant food by weeds may be even more injurious than the consumption of water. Only when water and fertility are far in excess of the needs of the crop could weeds do no harm.

The effect of witch grass in reducing yield is illustrated by data obtained at the New Hampshire station (Bulletin 71, page 55). Two plats of corn were treated in the same manner and given good cultivation up to June 10. One plat was hand-hoed four times after this date in order to destroy the witch grass, while this was allowed to grow in the other plat.

TABLE XLIX

EFFECT OF WITCH GRASS IN CORN

KIND OF CULTURE	POUNDS OF CORN STOVER	BUSHELS OF SHELLED CORN PER ACRE
Hoed	11,843	81.6
Unhoed	9,188	61.4

We may conclude that, for corn, the principal object of intertillage is to destroy weeds, and after this is accomplished, further tillage will not pay.

The above does not apply to small tilled crops, as vegetables where the soil is exposed and the roots do not fully occupy the surface soil. Here conditions approach those obtaining on fallow soil.

DEPTH AND FREQUENCY OF CULTIVATION

145. Since intertillage in corn apparently serves no important function beyond subduing weeds, it is to be expected that no increase in yield will result from cultivating more deeply or more frequently than is necessary in order to accomplish this purpose.

In TableXLVIII are shown results at the New Hampshire, Illinois, and Utah stations with deep and shallow tillage. The Illinois[1] results with methods of cultivation may be summarized as follows: —

TABLE L

Kind of Cultivation	Average Yield for Five Years Bushels per Acre
Frequent (4 plats)	68.6
Ordinary (4 plats)	68.5
Shallow (4 plats)	71.5
Deep (4 plats)	65.6

The principal injury of deep cultivation is that roots are destroyed. The depth to which the soil can be stirred without injury to roots depends on the soil to some extent. (See page 28.) In humid regions and clay soils, perhaps 2 inches is the limit; in loose loam soils in drier regions, the roots are ordinarily 3 inches below the surface; while with listed corn, the cultivation may often be as deep as 4 inches. The roots are usually shallow next to the plant and deeper midway between rows.

It is doubtful whether it would be an advantage to give deep culture, even when it could be done without particular harm to the roots, as illustrated with listed corn at the Kansas station.

Roots of listed corn are deeper than surface planted corn, and there would be little injury from deep cultivation.

[1] Ill. Agr. Exp. Sta., Bul. *31*: 356.

TABLE LI

RESULTS AT KANSAS STATION WITH DEEP AND SHALLOW CULTURE FOR CORN. AVERAGE FOR FOUR YEARS (1892–1896).[1]

TREATMENT	AVERAGE YIELD BUSHELS PER ACRE
Listed, deep culture	29.7
Listed, shallow culture	29.3
Surface planted, deep culture	27.3
Surface planted, shallow culture	27.0
Surface planted, deep and shallow culture[2]	28.1
Surface planted, surface culture	23.0

In Table XLVIII are also given results with frequency of cultivation. The following data from the Kansas station further illustrate:[3] —

TABLE LII

TIMES CULTIVATED	TIMES CULTIVATED TWO-YEAR AVERAGE	TWO-YEAR AVERAGE YIELD IN BUSHELS PER ACRE
Three times a week	17	24.8
Twice a week	13	27.2
Once a week	7	27.8
Once in two weeks	4	25.2
Once in three weeks	3	24.0
Once in four weeks	2	16.9

We may therefore conclude, from the data presented, that up to the time when corn shades the ground, and the

[1] Kansas Bul. *64*: 233.

[2] Deep first cultivation and shallow later.

[3] Kans. Agr. Exp. Sta., Bul. *45*: 131.

field is comparatively fallow, cultivation conserves some moisture as in any fallow soil. After the corn crop is thoroughly established and a layer of surface roots intercepts capillary moisture from below, the principal service of cultivation is to destroy weeds. Weeds compete with the plant for both water and plant food.

GROWING CORN FOR SILAGE

146. The general discussion has thus far had in view the culture of corn for grain. The recommendations taken as a whole apply quite as well to growing silage corn. It is generally true that the best quality of silage is made from corn grown under conditions for producing the maximum grain crop.

For grain it is necessary that the variety chosen should mature sound grain, but in the case of silage corn it need not mature. In the Southern States, and in practically all the Corn Belt States, perhaps the best silage variety is also the best standard variety grown for grain. In New England and on higher elevations in all Northeastern States, the most profitable silage variety will probably be too late to mature. At elevations of 1000 feet or more, seed may be secured at the same latitude but grown 500 to 1000 feet lower elevation. The growing season of corn usually shortens about one day to each 100 feet increase of elevation. At lower elevations it will be necessary to go 200 to 300 miles south for late seed. Dent corns are usually preferred for silage, Leaming being perhaps the most popular dent variety for this purpose. At higher elevations very early dents, sweet corns, and in some cases flint corns, are best.

As pointed out heretofore (page 179), the total weight of dry matter increases with rate of planting, but the proportion of ear decreases. In general, the best rate, yield and quality both considered, is about one-fourth to one-third thicker than would be necessary to secure maximum yield of grain under the same conditions.

Drills are best where the corn is planted somewhat thickly, as for silage. Even where hill planting has been found best for grain growing, drill planting has usually given slightly larger yields of stover. The difference, however, is too small to be of much importance, and the method to be adopted is to be determined by convenience in tillage and harvesting. Where harvesting is by machinery, drill planting is most convenient; but where harvesting is by hand, hills are preferred.

CHAPTER XVIII

ANIMAL AND INSECT ENEMIES

THE corn crop is more easily protected from its animal and insect enemies than most of the important crops. Of those insects that live on the roots of corn, practically all are effectively controlled by rotation. At present the corn rootworm and root-louse do considerable damage throughout the corn-belt, wherever several corn crops are grown in succession on the same land.

Rodents and birds do some damage every year, but are only considered serious, where corn is grown in small areas. The corn ear worm is difficult to control, but this insect seldom does serious damage except in the Southern States.

BIRDS

147. Crows give some trouble in regions where they are plentiful and the acreage of corn is comparatively small. They pull up the plants for a period of two weeks after the shoots appear, in order to get the kernels for food. Scarecrows or strings stretched with pieces of paper attached are effective in small fields. Coating the seed with coal tar is a deterrent, but not a complete preventive. The treatment consists in dipping a paddle in hot coal tar and stirring in the seed corn until every seed is coated with tar. The seed is allowed to dry and is then planted.

RODENTS

148. Small ground squirrels of several varieties dig up seed one to two weeks after planting. The coal-tar treatment recommended for crows is often effective as a preventive. Poison is also used. The ordinary method of poisoning is to soak a quantity of corn in a strychnine solution and plant this a few days ahead of the regular planting, in parts of the field likely to be molested. Very often the squirrels come mostly from adjacent pastures or meadows, and a few rows of poisoned corn planted next to these will be effective.

INSECTS

149. The larvæ of several insects are very injurious to corn under certain conditions. These may be grouped as: (1) Insects injurious to the roots. (2) Insects injurious to the young plant above ground. (3) Insects injurious to some part of the mature plant, as ear or leaf. (4) Insects that become abundant in cornfields only when corn follows corn year after year, as the corn rootworm. The remedy for this kind is rotation of corn with other crops. (5) There is another group, which injures corn only when it follows certain other crops. This includes the wireworm, which is often injurious the first and second years after grass sod. The grubworm is most often injurious after a clover sod. (6) Certain migratory insects, as the chinch bug, army worm, and stalk borer, which come in mostly from adjacent fields. The most important of these insects from an economic standpoint are here given, together with suggestions for their control: —

Cutworms

Cutworms live on various kinds of grasses. The moths lay their eggs in late summer. These eggs soon hatch and the partially grown larvæ live over winter in the ground. They live on vegetation again the following year and pupate during May and June. The larvæ feed principally during the night, cutting the young plants off near the ground. Late fall plowing usually destroys many of the larvæ. Late planting will often avoid them, and when the regular planting is destroyed it is usually safe to depend on a late replanting to escape. Cutworms are poisoned by mixing one pound of paris green to forty pounds of bran. When applied with a drill the mass is moistened and dried, so as to cause the poison to adhere. When applied by hand, a quart of molasses is added to the mixture.

Grubworms

These are larvæ of the May beetles, or June bugs. The eggs are laid in June, mostly in grasslands, but more or less in all cultivated fields, especially if recently dressed with barnyard manure. The larvæ live on decaying vegetable matter or roots, and often prove very destructive in cornfields.

No effective remedy has been proposed except in regions where listing is practiced. Listed corn is not injured so much as is surface-planted corn.

Wireworms

These are the larvæ of the family known as "click beetles." The eggs are laid in the spring, in soil on grassland. The larvæ usually live two years in the soil, then pupate in July and August, and are finally transformed into beetles in about four weeks. The larvæ both eat and bore the stems and roots of plants. No successful remedy has been proposed. When damage is expected, the corn may be planted more thickly, depending on thinning where the wireworms do not reduce the stand. When replanting a field injured by wireworms the new rows are planted midway between the old, leaving the old plants as food for the worms.

NOTE. The above pests, cutworms, grubs, and wireworms, give most trouble on grass sod. They seldom give trouble after cultivated crops where clean culture has been practiced.

There are two insects that are most troublesome where continuous corn culture is practiced — the corn rootworm and the root-louse.

Corn rootworm

There are two species, known as the Western and the Southern corn rootworm. The larvæ are similar and work in the same way, though the beetles differ in color. In early fall the female beetles lay about a dozen eggs in the ground near the corn roots. These remain over winter and hatch the next spring. The larvæ are about the size of a pin and two-fifths inch in length, almost colorless except for the head, which is yellow. They do most harm in July and August. Starting near the tip of a large root they bore inside the root, toward the plant. As they multiply rather slowly and as corn is their only host plant, the rootworms are serious only where the land has been in continuous corn culture for three or more years in succession.

Corn root-louse

Injury from the corn root-louse is very irregular, due no doubt to its natural enemies which ordinarily keep it in check. When unrestrained, however, it increases so rapidly that it may become very injurious in a short time. Usually its injury occurs in spots rather than over the whole field, due probably to local centers of infection from which it spreads rapidly. During the summer the wingless females produce living young continuously, which in turn at the end of a few days also begin producing young. The lice live on the juices that they suck from the corn roots. Winged females occur occasionally, which establish new colonies. In the fall both winged males and females appear. This last brood lays eggs which live over winter. Ants are often associated with plant lice and it is thought that they assist in protecting them and in caring for the eggs.

No practical way of restraining the lice has been suggested, except that early plowing and clean, thorough preparation of the land will destroy to a large degree those present in the soil.

The corn ear worm

The ear worm is the larva of a moth. Two to seven broods are produced each year, depending on latitude, about four broods being the average at the 40th parallel. It is the brood produced at silking time that is most injurious. The worms eat off the grains near the tip of the ear, not only destroying directly considerable grain, but also opening a way for fungous diseases and ear rot.

Migratory insects

Chinch bugs. — While chinch bugs breed in cornfields, the principal damage is due to migrating bugs from adjacent grainfields after harvest. The migration of wingless bugs is prevented by barriers, such as a dust mulch 10 feet wide, harrowed every day to keep loose, or a plow furrow with post holes every 2 rods where the bugs collect and may be destroyed by kerosene. A barrier of tar is sometimes used.

FIG. 73. — Ear of corn showing corn smut.

Army worms. — Where army worms migrate, the remedy generally recommended is to establish a post-hole barrier by plowing several furrows toward the colony; in the bottom of the last furrow, dig post holes into which the army worms fall and are killed with kerosene.

DISEASES OF CORN

150. The diseases affecting corn are the common corn smut (*Ustilago zea*) and certain ear rots, the most serious of which is caused by a fungus known botanically as *Diplodia zea*. Other forms of ear rot are caused by species of *Fusarium*. Both these diseases live over on infected stalks and ears, producing spores abundantly the following spring and summer to infect the new crop. The only remedy is to gather up and destroy by fire the infected material.

Corn is remarkably free from injurious diseases. It is rarely that the loss from smut or ear rot in a field will amount to so much as 1 per cent. Occasionally serious loss occurs.

References on insects injurious to corn: —

Ill. Agr. Exp. Sta., Bul. 44. Insect Injuries to the Seed and Root of Indian Corn. 1896.

Ill. Agr. Exp. Sta., Bul. 79. The Corn Bill-bugs in Illinois. 1902.

Ill. Agr. Exp. Sta., Bul. 95. The More Important Insect Injuries to Indian Corn. 1904.

Ill. Agr. Exp. Sta., Bul. 104. Field Experiments and Observation on Insects Injurious to Indian Corn. 1905.

Ill. Agr. Exp. Sta., Bul. 130. Experiments with Repellents against the Corn Root-aphis. 1905 and 1906.

Ill. Agr. Exp. Sta., Bul. 131. Habits and Behavior of the Cornfield Ant. 1908.

U. S. Dept. Agr., Farmers' Bul. 259. Corn Bill-bugs and Root Louse.

N. C. Agr. Exp. Sta., Bul. 203. Corn Weevils and Other Grain Insects.

Ky. Agr. Exp. Sta., Bul. 145. Corn Pests.

Ala. Agr. Exp. Sta., Circ. 8. Budworms in Corn.

U. S. Dept. Agr., Bur. Ent. Bul. 85. The Corn Root Aphis and Seed Corn Ground Weevil.

References of corn diseases: —

Kans. Agr. Exp. Sta., Bul. 23. Corn Smut.

Nebr. Agr. Exp. Sta., Bul. 11. Smut of Indian Corn.

U. S. Dept. Agr., Farmers' Bul. 69. Corn Smut.

U. S. Dept. Agr., Farmers' Bul. 234. Dry Rot of Corn.

Ill. Agr. Exp. Sta., Bul. 133. Ear rots of Corn.

CHAPTER XIX

HARVESTING THE CORN CROP

151. In the New England States, where corn culture first developed, it was the custom from the beginning to harvest the stalk as well as the ears. "Topping" was a common practice, the stalk above the ear being cut off for forage when immature, and later, when the ears had matured, these being "snapped" off and stored in barns to be "husked."

With the opening up of the North Central and Western States, from 1840 to the present time, corn became an important article of commerce. The acreage of corn increased rapidly and, with little use for the stover, the custom of harvesting only the ears became general.

In the Southern States, the corn area has never been extensive and a part of the forage has generally been saved. The custom of "topping" and "stripping" has been more general in the Gulf States than in other regions.

Corn has also been found to be the cheapest and best crop for silage; in dairy regions throughout the Northeastern States, corn is grown principally for silage, the entire crop of large dairy regions being utilized in this way.

In the Central and Western States, only a small proportion of the stalks are harvested for either silage or stover, but the practice of harvesting the entire plant is increasing. It is customary, when only the ears are harvested,

Fig. 74. — Typical cornfields in the corn belt.

to turn the farm live stock into the fields during the winter months to eat what they will of the leaves, husks, and smaller parts of the stalk.

TIME OF HARVESTING

152. The object should be to harvest at such a time as to secure the maximum amount of digestible food. The total dry weight continues to increase up to the time of ripening, as shown by the following data: —

TABLE LIII

Increase of Dry Weight as reported by Three Stations

Condition when Harvested	Approximate Date	Yield of Dry Matter per Acre				
		New York[1] (Geneva) Pounds	Michigan[2] Pounds	Kansas[3] Pounds	Average Pounds	Percentage of Increase
Ears in silk .	Aug. 10–15	3,000	3,670		3,335	
Ears in milk	Aug. 25	4,300	5,320	6,868	5,496	65
Ears in glazing . . .	Sept. 15	7,200	7,110	7,716	7,342	33
Ears ripe .	Sept. 25	8,000	8,020	9,548	8,523	16

[1] Ann. Rpt. 1889. [2] U. S. Dept. Agr., Farmers' Bul. *97:* 12.
[3] Kans. Agr. Exp. Sta., Bul. *30 :* 181–207.

At the time when corn is in tassel or in silk, less than one-half the dry weight has been developed. Increase in dry weight continues up to maturity. There was an average increase of 16 per cent from the time corn was glazed to time of maturity. There is an increase not only in total dry weight, but in all valuable constituents, as shown by the following data from the Michigan station —

TABLE LIV

YIELD PER ACRE OF GREEN FODDER, DRY MATTER, AND NUTRIENTS

TIME OF CUTTING	GREEN FODDER	DRY MATTER	PROTEIN	NITROGEN-FREE EXTRACT	FAT	FIBER
August 10 (tasseled)	21,203	3,670	472.7	1,828	67.9	1,010
August 25 (in milk)	25,493	5,320	576.0	3,212	143.1	1,148
September 6 (glazing)	25,865	7,110	711.0	4,554	199.0	1,294
September 15 (ripe)	23,007	8,020	696.9	5,356	242.6	1,413

Not only does the total yield increase, but the quality improves with maturity. The large group of compounds under the head "nitrogen-free extract" are not all equally valuable for feeding purposes. Starch and the sugars are the most valuable and both increase in proportion as the plant matures, due to the development of ear, as shown by Jordan of the Maine station.[1, 2]

TABLE LV

	PERCENTAGE OF STARCH AND SUGAR IN NITROGEN-FREE EXTRACT	POUNDS OF STARCH AND SUGAR PRODUCED PER ACRE
August 15, ears beginning to form	25.1	358.5
August 28, a few roasting ears	40.5	1,172
September 4, all roasting ears	42.7	1,545
September 12, some ears glazing	42.2	1,764
September 21, all ears glazed	50.3	2,244

[1] Maine Agr. Exp. Sta., Bul. *17*: 4.

[2] U. S. Dept. Agr., Farmers' Bul. *97* : 12.

RELATIVE PROPORTION OF PARTS

153. Before considering the time and method of harvesting the whole plant, it will be well to note the relative proportion and value of the different parts of the corn plant at various stages of growth. The Michigan station has studied this subject and reported the following results:[1] —

TABLE LVI

Percentage of Total Dry Matter in Leaves, Stalks, and Ears of Corn Plants at Four Stages of Growth (Michigan Station, 1896)

Time of Cutting	Percentage of Total Dry Matter		
	Leaves	Stalks	Ears
August 24 (in milk)	36.41	34.27	29.32
August 31	33.63	25.52	40.85
September 7 (glazing) . . .	30.03	25.53	44.44
September 14 (ripe) . . .	21.77	31.91	46.32

COMPOSITION OF PARTS

154. The total dry weight alone does not give a comparative statement of the relative feeding value of the parts of a corn plant. The leaves are very high in albuminoids, while the stalks are low in these compounds. Pound for pound, leaves are about twice as valuable as stalks. A further study of the distribution of the principal compounds of the plant at different stages of growth is reported as follows: —

[1] U. S. Dept. Agr., Farmers' Bul. *97:* 9–12.

TABLE LVII

DISTRIBUTION OF ALBUMINOIDS AND NITROGEN-FREE EXTRACT IN LEAVES, STALKS, AND EARS OF CORN AT DIFFERENT STAGES OF GROWTH

TIME OF CUTTING	ALBUMINOIDS			NITROGEN-FREE EXTRACT		
	Leaves	Stalks	Ears	Leaves	Stalks	Ears
August 24 (in milk) .	52.50	10.00	37.50	38.50	17.50	44.00
August 31	51.06	2.53	46.41	28.40	23.64	47.96
September 7 (glazing) .	42.71	5.19	52.10	20.50	25.30	54.20
September 14 (ripe) .	30.60	10.70	58.70	15.90	29.40	54.70

The above tables show very clearly the shift in relative proportion of dry weight and important food constituents from leaves and stalk to ear, as growth progresses. From the data presented in the last five tables it would seem that corn should be allowed to stand until quite mature before harvesting, since the total yield and quality apparently improve. There are two considerations against this: the loss of leaves, and the fact that both leaves and stalk become less palatable with maturity.

RELATIVE VALUE OF PARTS

155. From the last two tables it appears that at the time the ear is in the "milk" stage, the relative dry matter is about equally distributed between leaves, stalks, and ears, although 40 to 50 per cent of the total nutrients are in the leaves alone. There is then a gain in ear until 46 per cent of the dry weight and about 56 per cent of the nutrients are found in the ear.

RELATIVE FOOD VALUE OF EARS AND STOVER

At the time corn would be cut for silage or fodder, when the ears are glazed, about 40 per cent of the protein and 20 per cent of the nitrogen-free extract are in the leaves; or, of the total food value of the plant at this time, approximately 30 per cent is in the leaves, 15 per cent in the stalk, and 55 per cent in the ear.

Armsby[1] compiled the data from four stations and calculated the yield of ears and stover to be as follows: —

TABLE LVIII

Station	Ears	Stover
New Jersey (dent)	4,774	4,041
Connecticut (flint)	4,216	4,360
Wisconsin (dent)	4,941	4,490
Pennsylvania (dent)	3,727	2,460
Average	4,415	3,838

The above average shows that about 53 per cent of the crop by weight is ears; but the ears contain a higher percentage of digestible nutrients than does the stover, and a calculation of the digestible nutrients in the above shows about 63 per cent in the ear and 37 per cent in the stover. The above figures represent the distribution of nutrients at the time the stover is cut for forage, but do not indicate the final distribution of digestible nutrients. Fodder is usually cut when the ears are glazed in order to save the valuable leaves, and about ten days before it is ripe. But during this period there is considerable translocation of sugars and starch from the leaves and stem to

[1] Penn. Agr. Exp. Sta., Rpt. *1887*.

the ear, so that in the fully matured corn crop, under normal conditions, between 60 and 70 per cent of the digestible nutrients will be in the ears.

This ratio would not apply to corn planted thick for silage, when the proportion of stover is increased without decreasing the yield of ears.

There is also considerable increase in total weight between the time the ears are glazed and the time when they are ripe, usually amounting to about 10 per cent. The value of stover obtained must be decreased by whatever loss is occasioned by early harvesting. Charging this loss against the stover, it would appear that the total feeding value of the crop is increased about 25 per cent by harvesting the stover when the ears are glazed, in comparison with allowing the crop to mature and harvesting only the ears.

In conclusion, corn should be permitted to become as nearly mature before harvesting as is practicable. As pointed out heretofore (page 227), two-thirds of the value of the stover is in the leaves, and it is therefore important to save these. In a humid climate, with fall rains, it is often possible to allow corn to stand until most of the ears are mature before cutting; but in a region with dry falls and windy weather the harvesting must be done seven to ten days earlier, if the leaves are to be saved.

TIME OF HARVESTING FOR SILAGE

156. When the silo first came into use, the custom was to use very immature material. It was found in time that silage from mature corn was better in quality and the yield was greater. There is a limit, however, in this direction. Silage, in order to keep well, must pack closely, and as nearly as possible, all air must be excluded. Corn too mature cannot be packed closely enough, though sprinkling with water and careful tramping will allow the ensilaging of corn even when more than half the ears might be considered ripe. As a general rule, when the husks have mostly turned yellow, and two to four bottom leaves have turned, is the proper time.

Fig. 75. — A modern silage cutter, with blower attachment, for delivering the cut silage.

Good silage contains about 75 per cent water, and it is doubtful whether it would be practicable to ensile corn containing less than 65 per cent moisture.

METHODS OF HARVESTING

157. The four methods of harvesting maize are as follows: —

1. Stripping: leaves removed while green for forage, and ears husked later.

2. Topping: tops cut off above ear for forage, and ears husked later.

3. Ears only harvested, stalks left in field.

4. Entire plant harvested for silage or fodder.

Harvesting by hand

158. Stripping and topping are practiced in the belief that in this way the forage may be obtained while green and in the right condition to harvest, while the ears are allowed to remain and mature. It has been

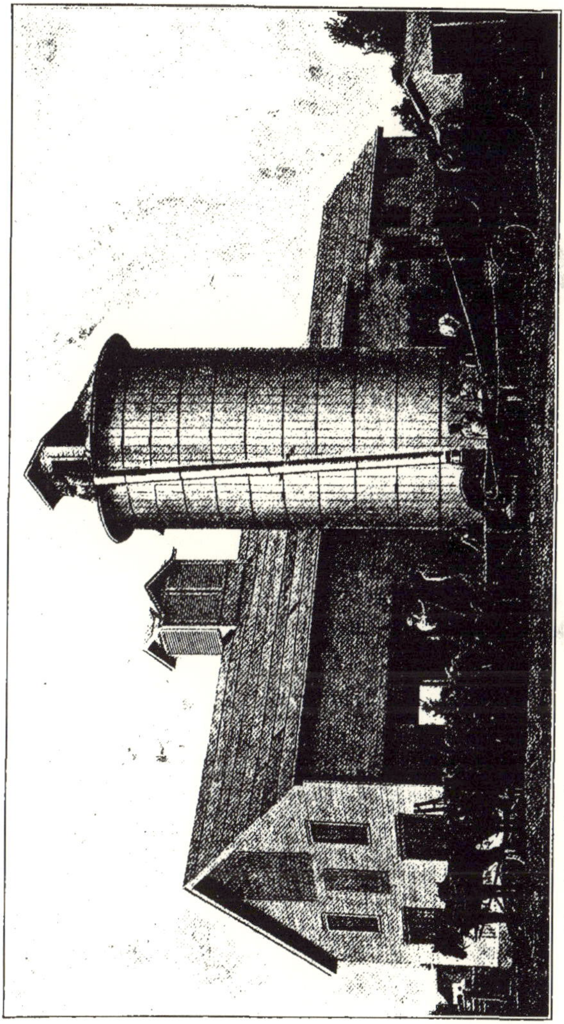

FIG. 76. — A silo and silage cutter in operation. Corn is the principal crop used for silage.

shown,[1] however, that both stripping and topping reduce the yield of grain, so that it is doubtful whether the total yield of grain secured is greater than when the whole plant is harvested as fodder. The loss of shelled corn has generally amounted to 10 to 20 per cent, which is about the usual loss when harvested as fodder.

The Texas station reports the labor expense of topping and stripping to be as follows: —

Tops only: Cost per ton of dry-cured fodder . . .	$2.13
Leaves only: Cost per ton of dry-cured fodder . .	7.67

As it takes about four acres to produce a ton of leaves and half as much for a ton of tops, the value of the forage secured does not compensate for the loss of grain and cost of harvesting.

159. Hand cutters. — Probably the first tool used in harvesting fodder was the hoe. Corn knives came into use in time, those made from old scythe blades being the most common at first. Corn "hooks" were also made by inserting a short blade at about right angles in a short wooden handle. There are several standard types of knives and hooks on the market.

Horse-drawn cutters

160. The first horse-drawn cutters to have a general use were sleds, drawn astride of the corn row, with a heavy knife attached in front at the right height to cut off the corn plants, or drawn between two corn rows with a heavy knife attached to one or both sides for cutting

[1] Miss. Agr. Exp. Sta., Bul. *33:* 63. 1895.
Penn. Agr. Sta., Rpt. *1891:* 58–60.
Ga. Agr. Exp. Sta., *23:* 81–82. 1893.
Ark. Agr. Exp. Sta., Bul. *24:* 120.

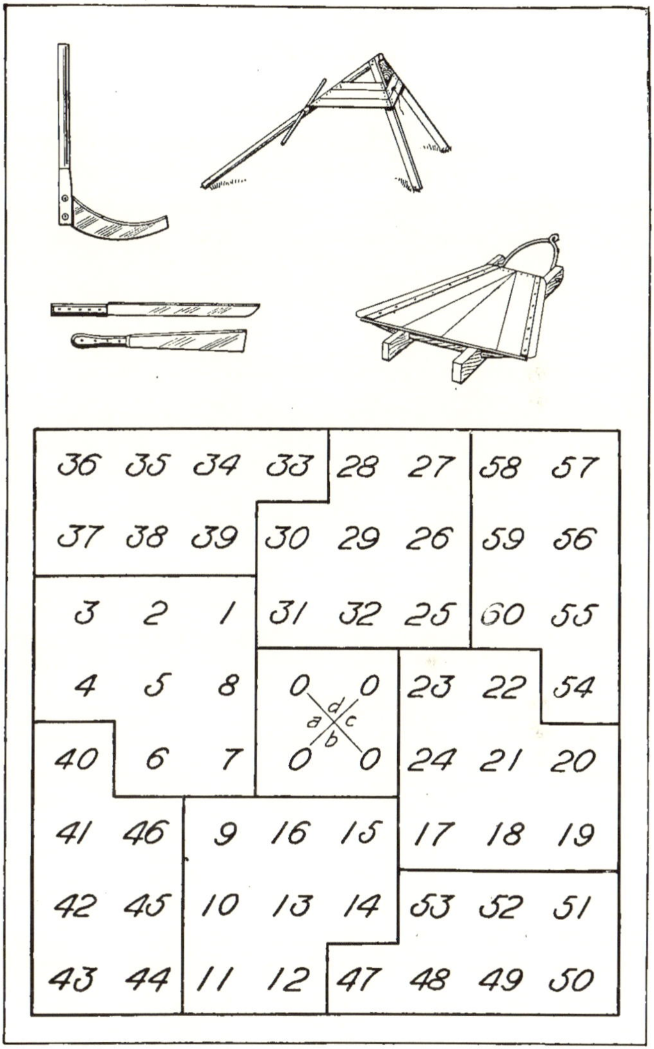

FIG. 77.—A corn hook and knives used in harvesting corn fodder. A "horse" used in shocking corn fodder. A home-made sled cutter. The sled is drawn by a horse between two rows, the stalks being cut by sharp knives on each side. Two operators stand on the sled.

The lower figure illustrates a system of cutting by hand, in order to economize steps.

the plants. Later, wheels were substituted for runners and seats were provided for the men. Some cutters have large platforms to carry the green fodder until enough has been accumulated for a shock. There is no labor saved

FIG. 78. — A two-row corn cutter mounted on wheels. The two operators stand between the wheels.

by having a large platform, and the most popular type is that in which there is room for only a large armful to be collected at a time.

The corn binder

161. The first successful corn binders were introduced about 1895 and have steadily increased in popularity. The corn is bound in bundles of convenient size, and with a bundle carrier six to eight bundles may be collected before dropping windrows to be shocked up later or drawn to the silo.

Shocking corn

162. The ordinary custom in curing fodder is to leave it in shocks for one to three months. It is then sufficiently

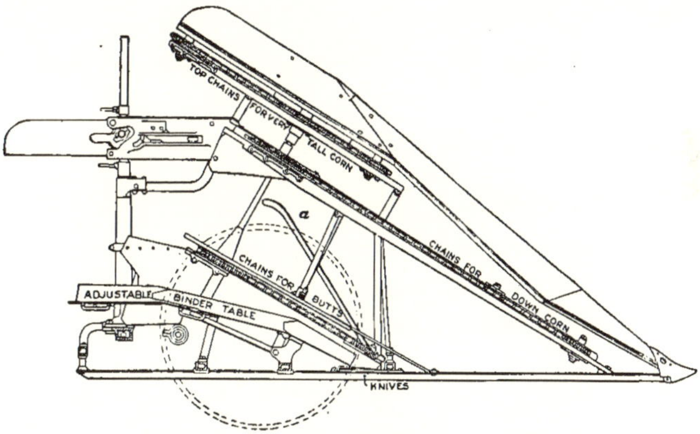

FIG. 79. — Corn fodder harvester in section.

cured to husk or store in barns or stack yard. It is often left in the field to be hauled as needed during the winter.

Size of shocks

163. The exposure and loss is greater in small shocks than in large. Where fodder is green, the shocks must be small if the corn is set directly into shock, ordinarily one hundred to one hundred fifty hills being enough. When cured it is often practicable to set two or three shocks together or to stack. When the fodder can be allowed to partly cure before shocking, as in harvesting with a binder, the shocks should be made as large as is practicable.

FIG. 80. — The corn harvester and binder in operation.

Setting up shocks

164. When cutting corn with knives, it is customary to tie four hills together for a " horse " in the place where it is proposed to place a shock. In other cases a " horse " is made as illustrated in Fig. 77. In setting up bundles after a corn binder, a " horse " is not necessary.

Tying shocks

165. After the shock is well set up, the tops of the outside stalks should be tucked under and the shock securely tied with binder twine. A rope with iron hook on one end, or a quirt, is useful in drawing the shock before tying.

When corn is cut by hand, some steps will be saved by following a systematic plan, in cutting the hills for each armful. Such a plan for a shock ten hills square is illustrated in Fig. 77.

Husking fodder corn

166. The fodder may be husked in the field, a common practice in the West, or as common in the East, hauled to the barn to be husked later, or hauled to a shredder. The shredder delivers the shredded fodder and husked ears in separate piles. When husking by hand in the field the ears are often thrown into piles, to be collected later with a wagon. A more convenient way is to husk directly into the wagon. A high " throwboard " should be put on the wagon box opposite the husker. A light frame on wheels may be attached to the rear of the wagon across which the fodder corn is thrown for husking. This allows the husker to stand while at work.

Shredding fodder

167. Zintheo makes the following statement:[1] "Between 1880 and 1890, a great deal of attention was given to threshing corn. This practice so battered the stalk as to make every part of it available as a cattle food. Fodder cutters had been in use for many years yet this method of preparing corn fodder left the fibrous part of the stalk in a tough woody condition which cattle did not relish. The bruising and shredding action of the thresher put the stalk in a more palatable form. The repeated shortages and failures of the hay crop during the decade 1880–1890,

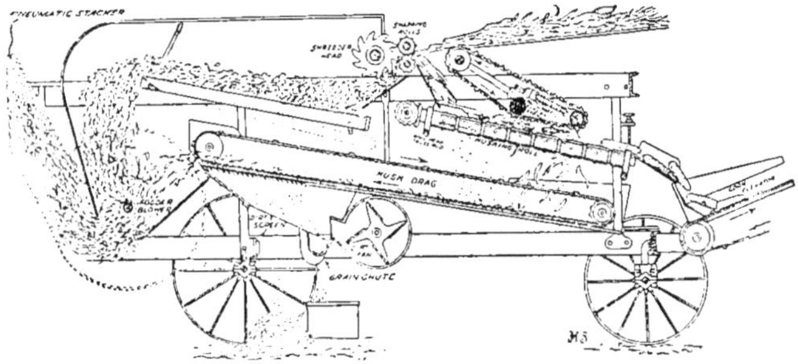

Fig. 81. — Combined shredder and husker.

together with the results of attempts at threshing corn, led to the invention of the combined husker and shredder, which takes the stalks with the ears on them and prepares the stalks for feeding."

Shredding fodder is generally considered as an economic way of preparing corn fodder for feed. In humid climates there is sometimes trouble with the shredded fodder heating when piled in large quantities, unless care is taken to shred only fodder in a fairly dry condition.

[1] U. S. Dept. Agr., Office Exp. Sta., Bul. *173*: 40.

Hauling fodder corn

168. When there is snow, a sled with fodder rack is most convenient. At other times and for drawing silage, a low down rack on wheels is desirable.

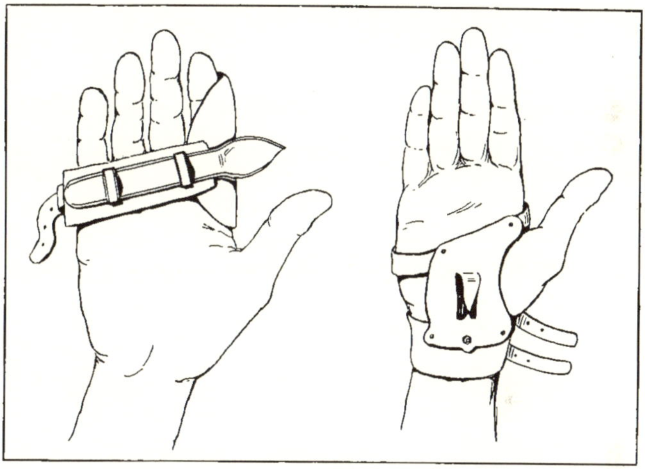

Fig. 82. — Husking peg and husking hook. The peg is best for fodder corn and the hook for standing corn.

Harvesting ears by hand

169. In the Corn Belt States, only the ears are harvested on perhaps nine-tenths of the area. The method is to husk directly into a wagon. A "throw-board" about 30 inches high is put on the wagon-box on the far side from the husker. The husker takes two rows at a time and usually one man to a wagon. An average day's husking in good corn is 60 to 75 bushels of shelled corn. The husker uses a peg or hook in the palm of his hand to assist in tearing off the husks.

Harvesting ears by machinery

170. For at least fifty years, attempts have been made to devise mechanical corn pickers to operate in the field. Within the past few years, machines have been perfected

FIG. 83. — Method of husking corn from the field in corn belt.

that do the work in a satisfactory manner, provided the stalks stand up well and too many ears have not fallen to ground. At best, some ears are left in the field, which must be picked up by hand. In some cases, live stock are turned in to gather up ears that are left. As the machine requires six horses to draw it and two more teams to draw the ears away, it is only practical in large fields. A machine will husk about eight acres a day.

Comparative cost of harvesting methods

171. Zintheo [1] has collected and summarized data on comparative cost of different methods of harvesting corn. He gives the following estimate as comparative cost in the corn-belt, where the corn is producing an average of 44 bushels per acre : —

TABLE LIX

COST OF HARVESTING CORN BY VARIOUS METHODS

Average data for harvesting by hand

Cost of implement	$1.00
Acres one man harvests per day	1.47
Cost of cutting and shocking	$1.50 per acre

Average data for harvesting with sled harvester

Cost of implement	$5 to $50
Acres 2 men and 1 horse harvest per day	4.67
Cost of cutting and shocking	$1.18 per acre

Average data for harvesting with corn binder

Cost of implement	$125.00
Acres cut per day by 1 man and 3 horses	7.73
Acres shocked per day, 1 man	3.31
Cost of cutting and shocking	$1.50 per acre

[1] ZINTHEO. Corn Harvesting Machinery. U. S. Dept. Agr., Office of Exp. Sta., Bul. *173:* 46.

Cost per bushel of picking and husking corn

	CENTS
By hand from field	3.5
Team for cribbing	1.
By hand from shock	5.3
Team for cribbing	.79
By corn picker from field	4.1
By huskers and shredder from shock . .	4.5

The relative cost of methods will differ, depending principally upon the price of labor.

Storing ears

172. The ears are usually stored in slatted cribs to provide ventilation. If a good roof is provided, there is

FIG. 84. — A good type of farm corn crib, and farm elevator used in unloading.

seldom loss from rotting in the crib. Rats and mice cause considerable loss where corn is stored for several months or more and it is important to have cribs rodent proof. Ventilated sheet-iron cribs are now on the market, that are rodent-proof if set on a cement foundation. Wooden cribs can be made rodent-proof by lining with hardware netting or if constructed on a cement foundation. Where a cement floor and foundation are used, care must be taken to provide ventilation underneath by means of a raised board floor. The floor may be slated and made in movable sections to facilitate cleaning beneath.

Shrinkage in curing fodder and silage

173. If the total dry matter and protein content of corn fodder be ascertained at the time of storage either as fodder or silage, it may be determined that there is a constant loss in both for at least a year. The amount of this loss as determined by the Wisconsin station is summarized as follows: —

TABLE LX

LOSS IN CURING CORN IN SILO OR AS FODDER (WISCONSIN STATION, THREE-YEAR AVERAGE)

METHOD	GREEN FODDER POUNDS	SILAGE OR DRY FODDER POUNDS	LOSS IN POUNDS	CURING PER CENT
(*a*) Ensilage method				
Dry matter . . .	35,602	28,300	7,281	20.5
Crude protein . . .	2,910	2,312	597	20.6
(*b*) Field cured				
Dry Matter . . .	39,448	31,428	8,020	20 3
Crude protein . . .	3,102	2,619	482	15.6

The Connecticut station [1] reports results of an experiment in which no loss was apparent while curing. Most

[1] Conn. Sta., Rpt. *1889* : 219.

experiments, however, show field losses ranging from 10 to 20 per cent. A part of this loss in field curing is due to direct loss of leaves and portions of the stalk. Where direct loss of material is entirely prevented, there is still a loss, apparently due to a slow process of oxidation or fermentation. This loss will go on even when placed in stack or under cover.

As 15 to 20 per cent of the feeding value of corn fodder is in the leaves, a large share of the loss of field curing is due to loss of leaves, but a part to fermentations; on the other hand, all loss in silos is due to fermentations.

Gain in gross weight

174. After fodder has become thoroughly air dry, its weight will then vary with the humidity of the air, as dry fodder readily absorbs moisture. The Connecticut station reports the results with two lots of fodder in 1877. The fodder-crop was very heavy, but the fall being dry, the two lots cured down to 27 per cent and 36 per cent moisture respectively, when placed in the barn. The winter was warm and damp, so that 5.2 tons placed in the barn Nov. 11, had increased in weight to 8.5 tons by Feb. 8.

Shrinkage of ear corn in storage

175. When ear corn is stored as harvested in October or November, there is a shrinkage in total weight during the first year varying from 5 to 20 per cent. Shrinkage is principally due to drying out of water. It is directly related to how well the corn matures, and the dryness of fall weather. The following data from three experiment stations illustrate; —

TABLE LXI

SHRINKAGE OF CORN IN CRIB AS SUMMARIZED FROM RESULTS OF THREE STATIONS

MONTH AFTER HARVEST	KANSAS[1] THREE-YEAR AVERAGE PER CENT	ILLINOIS[2] FOUR TRIALS PER CENT	IOWA[3] ONE YEAR PER CENT
December		3.6	8.7
February	3.26		
March		5.7	10.5
April	5.16		
June	6.80	14.4	16.2
August	7.44		
September		16.6	19.4
October	8.62		

[1] Kans. Bul. *147* : 267. [2] Ill. Bul. *113* : 363. [3] Iowa Bul. *45* : 228.

Results at the Illinois station show practically no loss the second year. In fact after corn has become thoroughly air dry, the weight will then fluctuate with the humidity of the air, the variation amounting to as much as 3 per cent. (See Ill. Bul. 113, p. 363.)

Shrinkage is partly due to the loss of water, but as pointed out by Ten Eyck [1] the loss of moisture alone does not account for the entire decrease in weight. There is a decrease in actual dry matter probably due to some process of oxidation.

Marketing

176. Corn is usually marketed as shelled corn and is seldom shipped in ear. About 60 per cent of the corn crop is consumed on the farms where produced. About 10 per cent is sold locally to feeders and about 25 per cent

[1] Kans. Bul. *147* : 268.

finds its way into the general markets. Of the total crop produced in the United States, about 3 per cent is exported, hence a large share of the corn reaching the general market is redistributed in the United States.

The ear corn is usually stored on the farm in cribs holding 500 to 5000 bushels. When ready to market it is

FIG. 85. — Large power corn sheller in operation on a farm. Will shell 400 bu. per hour.

shelled out, with power shellers that handle 200 to 400 bushels per hour. The shelled grain is then hauled to a local elevator where it is loaded on cars and shipped either direct to a consumer, or to one of the large terminal elevators, where the grain may be stored.

Drying corn for shipment

177. Corn is comparatively easy to keep in storage, the principal difficulties coming from excess moisture. When corn is shipped in cars, from northern states to the south, or when loaded in ships for export, there is great danger that it will " go out of condition " if containing higher than 15 per cent moisture. Large commercial driers are now in general use, capable of drying several thousand bushels a day, to 12 per cent moisture.

Cost of producing

178. The principal cost factors in producing corn are labor, and rent of land. The cost of seed and fertilizer being minor factors, at present, in the corn-belt, although

FIG. 86. — Large cement grain tanks, such as are used for storage at terminal markets. (Erie Railway, Chicago, Ill.)

in certain sections of the East and South, the use of fertilizers on corn is becoming more common.

The rent of land is fairly well standardized, being in general about $5 per acre for land capable of producing 40 to 50 bushels per acre. The amount of labor varies with soil. The required labor to produce an acre of corn on the heavy clay lands of the East is probably twice that required on the prairie land of Iowa and would be still less in central Nebraska and Kansas, where listing is a general practice.

The cost of growing and harvesting ears from standing stalks has been reported from many sources, the general results being illustrated by the following data: —

TABLE LXII

	Date of Investigation	Cost per Acre of Raising Dollars	Cost of Harvesting Ears from Standing Stalks Dollars	Yield per Acre Bushels	Total Cost Dollars	Cost per Bushel Cents
American Agriculturist summarized from several states in the corn belt[1]	1897	8.43	1.00	39.2	9.43	24.0
Minnesota[2]	1902–04	8.25[4]	3.51	40[4]	11.76	29.4
		7.35	2.60	40[4]	9.95	24.9
Nebraska[3]	1909–10	10.06	1.59	39.3	11.62	29.6

[1] Book of Corn, p. [2] Bureau of Statistics, Bul. No. 48, p. 41.
[3] Nebr. Bul. 122, p. 9. [4] Estimated.

Earlier estimates when both land and labor were cheaper indicate that corn was produced for 20 cents per bushel in the period from 1885 to 1895. The fertility of land is an important factor in the cost per bushel or ton, as the expense of raising is little if any more on good land than poor.

The cost of harvesting fodder corn and silage has been estimated in another place (page 241).

CHAPTER XX

USES OF CORN

179. Perhaps nine-tenths of the corn crop is fed to live stock. The remainder is used in the arts, in manufacturing glucose, starch, corn meal, breakfast foods, hominy, corn oil, and alcohol, etc. The husks are used in matting, the stalks and pith in packing, and corn cobs are used in making tobacco pipes.

Corn meal and hominy have been important articles of food among American people from Colonial days. The use of corn as food has declined since the Civil War, probably due to the large production of wheat at low cost. The principal corn-meal market at present is in the Southern States, where it is extensively used by the people of both races. There is a general but light demand for " fancy corn-meal " throughout the country.

The two principal grades of meal are whole meal and " degerminated " meal. In the first case, the whole corn is ground and only the coarsest bran removed, giving a yield of about 94 pounds of meal from 100 pounds of corn. This meal contains all the germ which darkens the color and adds its own flavor. Within recent years, degerminating has become general in making fancy meal. The germ and bran are all removed, the meal well ground and bolted, giving about 40 pounds of meal to 100 pounds of corn. This meal is often called " granulated " meal.

Corn bran and germ meal are the by-products of meal manufacture, both of which are used for stock food, while the germ meal is also used in the manufacture of prepared breakfast foods.

Hominy is whole or cracked corn with the hull removed. Originally hominy was prepared by soaking the whole corn in a strong lye solution, which caused the hulls to loosen and was then removed by washing, but at present, the hulling of commercial hominy is done with machinery.

Grits is coarse ground hominy, but the commercial product is usually prepared as an intermediate product in the grinding of meal.

Germ meal is a by-product in the manufacture of corn-meal and starch and is composed principally of germs.

Glucose or corn sirup is made by inverting the starch of corn by means of dilute hydrochloric acid. The germ is first removed and put on the market as germ meal or pressed to extract the oil. *Gluten feed* is the residue after glucose is extracted and is very rich in protein compounds and has a standard market value as stock food.

Corn oil is extracted by pressure from the separated germs, which are about 30 per cent oil. The oil is used as a salad oil, in paints, or vulcanized as a substitute for vulcanized rubber. The residue after extracting oil is known as corn oil cake.

Starch. — Corn was an important source of starch at one time, but potatoes are more commonly used at present. The starch is extracted by washing from the corn flour. A residue is left known as *gluten feed.*

Distillery products are the residue left as a result of distilling alcoholic beverages. The starch is largely removed in distilling, leaving a iermented by-product, high in protein content, which is put upon the market in various forms as stock food.

Pop-corn products. — A large proportion of the pop-corn crop is utilized with no other preparation than popping, with a small amount of butter and salt added for seasoning. The popped corn is also used in various confections and in prepared breakfast foods.

Sweet corn products. — The sweet corn crop is utilized as green corn on the cob, as " canned " corn, and " dried " corn. *Dried corn* was at one time an important home-made article of food and considerable was sold as a commercial product. Canning is the principal method of preserving green corn and has become an important commercial industry.

Cereal food products. — Corn, either as grits, germs, or popped, is utilized to some extent in various prepared cereal foods. A common method is to cook the cracked hominy until soft, then roll into thin flakes, which are then dried. The cooking and drying increases the soluble sugars, and more or less carmelizes the carbohydrates.

Corn meal is utilized in various ways. Corn-meal mush and samp is made by simply boiling in water with a little salt, and is well known. *Polenta,* made in the same way, is said to be almost a national dish in Italy.

The three principal forms of bread made from corn meal are hoe-cake, johnny-cake, and brown bread ; the formulas for which are given on the following page.[1]

In the "Cotton Belt" of the United States corn furnishes the principal bread food, rather than wheat. In other sections of the United States — also in Europe — corn meal is used in proportions varying from 5 to 50 per cent in a variety of bread and pastry foods, where a coarse flour is desirable.

[1] Maine Bul. 131, p. 139.

	JOHNNY-CAKE GRAMS	BROWN BREAD GRAMS	HOE-CAKE GRAMS
Corn meal	100.0	100.0	100.0
Flour (wheat)	100.0	100.0	....
Salt	5.0	4.0	5.0
Sugar	10.0	....	5.0
Baking powder	4.4	4.4	....
Molasses	....	40.0	....
Water	....	....	400.0
Milk	150.0	200.0	....

References on corn as food · —

Food Value of Corn and Corn Products. Farmers' Bul. 298. 1907. Indian Corn as Food for Man. Maine Bul. 131. 1906.

CHAPTER XXI

SHOW CORN

THE culture and characteristics of show corn deserve special discussion.

180. Corn shows, in common with poultry and live stock shows, serve a practical purpose, in so far as they sustain

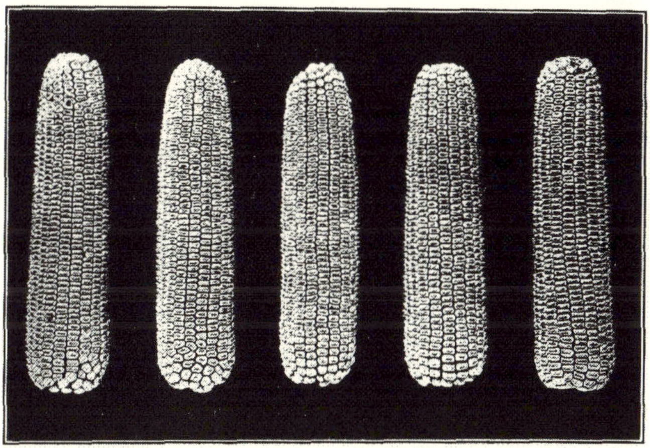

FIG. 87. — Show ears of Boone County White. A typical white variety of the corn-belt.

interest, and serve as a rallying ground for those interested in production.

On the other hand, show corn is not necessarily the best type to grow or most productive, and farmers have often

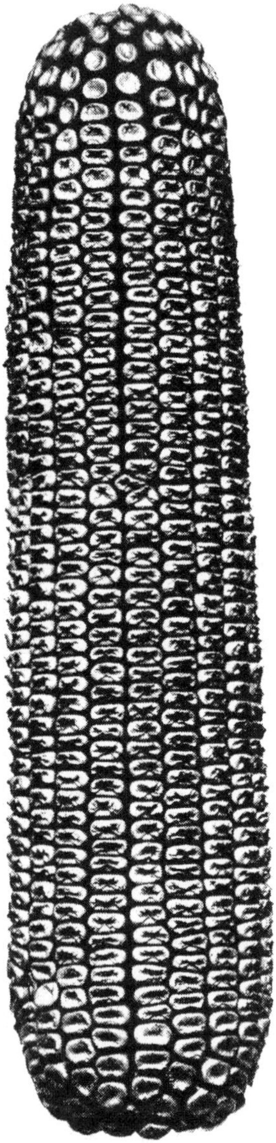

made the mistake of buying it to plant under conditions not suited to that type of corn.

Usually show corn is grown under the most favorable conditions of climate and soil. There are certain regions, such as southern Indiana, central Illinois, and Missouri River bottom land, where corn apparently attains a perfection in type not possible under average conditions.

Our study of acclimatization developed the importance of growing seed corn under conditions similar in soil and climate to the region where it is to be used as seed. This makes it doubtful whether corn grown under the most favorable environment is best adapted for average conditions.

The future of corn shows does not rest so much on practical considerations as æsthetic. A sound, perfect ear of corn is beautiful, artistic, and pleasing to the senses. The plant on which it grew is interesting in the same way.

FIG. 88. — A typical ear of show corn. Ried's yellow dent.

The ear also represents the largest and most interesting crop in the United States, and the principal means of support of many millions. So long as men admire perfect ears of corn, the corn show will last.

181. Show corn is judged, on the basis of degree of perfection exhibited, both in soundness and general symmetry, uniformity, and beauty. It must be perfectly sound and matured, and free from signs of deterioration due to disease or improper care.

The characters of show corn may be grouped in two classes, as those that pertain to soundness and maturity and those that pertain to perfection in symmetry and uniformity. The first class is of practical value and applies in the judging of all seed corn. The second class of points cannot be said to be important to consider in seed selection.

182. Maturity is judged by the general plumpness and development of the kernels. If the kernels are loose on the cob, or unduly shrunken at tip or crown, the ear probably did not mature properly.

183. Soundness is judged principally by the vitality of germs and strength of germination. Good germs should be plump, of a texture similar to good cheese, and no signs of discoloring. Any variation from this can usually be seen, but it is not always possible to judge the viability by examination alone. A germination test is sometimes necessary to determine this point.

Fancy characters pertain to the perfection and symmetry of development of all parts of the ear, as butts, tips, rows, kernels, etc.

184. Standards of perfection have been adopted in regard to a few of the best-known varieties, but at present these standards are not regarded very much by corn judges, but rather a universal standard has come to be recognized, which is applied to all exhibits, more or less regardless of variety.

For dent corn the following standards are generally accepted:

Shape of ear. — Cylindrical or nearly so. The circumference should be about three-fourths the length.

Size of ear. — The standard size of large dent varieties is ten inches in length and seven and one-half inches in

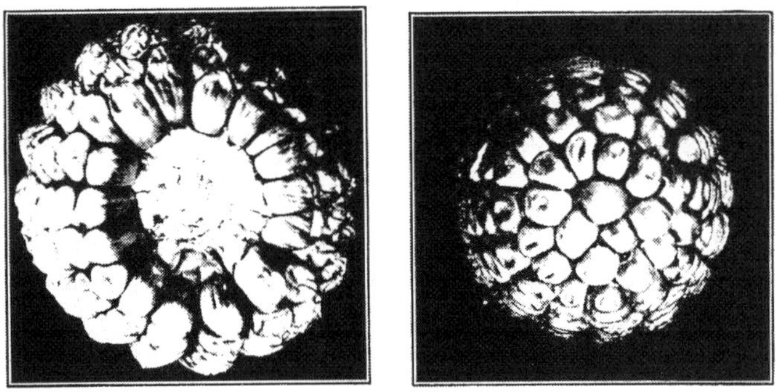

FIG. 89. — Ideal butt and tip ends of dent corn. Note the regular size of kernels in both cases.

circumference; of medium dents, eight inches long and six inches circumference.

Rows. — The rows should be straight, and each row be full length of ear and extend well over butt and tip. Short or irregular rows are regarded as imperfections.

Butt ends. — The butt end should be well rounded, not flat. The shank should be about one-half the diameter of cob. If smaller, the ear is liable to fall off the stalk; and if larger, the ear is more difficult to husk.

Tip of ears. — The rows should extend in a regular way well over the tip. Only a small exposure of cob at the tip end is allowed. Full depth of grain should extend almost to the very tip of the ear.

Type of kernel. — A good kernel of large dent corn should be about seven-eighths inch in length and three-eighths in width, if an eighteen-row ear, but narrower if more rows. The kernels should fit close from tip to crown, being somewhat keystone shaped. The kernels should be fairly thick, averaging in the row about six kernels to the inch. The kernel tip should be full and square; the germ, large, plump, and of good color and texture.

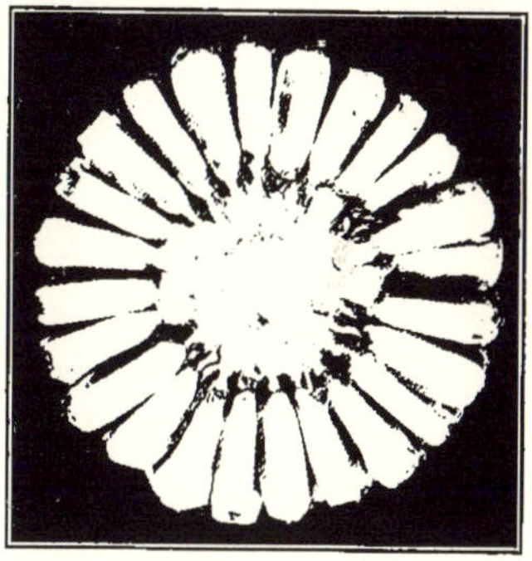

Fig. 90. — Cross-section of very deep-kerneled type of dent corn commonly known as hackberry.

GROWING SHOW CORN

185. The seed must come from a good show strain with many generations of selection for type. The soil should be naturally good corn soil, and everything done to put the soil in perfect condition, by proper rotation, manuring, and tillage. The soil, however, can be too rich in nitrogen for best results, as the plant is then inclined to run too much to stalk rather than ear. The soil should be rich in available minerals. Good show ears seldom come from the portion of field where the growth is rankest, but rather from a part where growth of stalk is normal but ears large.

The rate of planting should be rather thin, about two-thirds normal stand.

The crop should be handled so as to insure a rapid normal growth throughout the season without a check.

FIG. 91. — An example of prolific corn.

CHAPTER XXII

SWEET CORN OR SUGAR CORN

BY ALBERT E. WILKINSON

SWEET corn is grown chiefly as a vegetable for table use, although the stover is usually harvested as forage for stock. Sometimes sweet corn is planted as a silage or forage crop. The development of sweet corn has been discussed in another place (page 79).

VARIETIES AND TYPES

186. Sweet corn may be divided into about the same general classes and types as field corn. The height of the stalk varies from three to ten feet and the number of rows on the ear from eight to twenty. Practically all common colors are found. The time from planting to maturity varies from 65 to 110 days.

187. As mentioned before (page 23) sweet corn is any one of the starch corns (flint, dent, or flour corn) that has lost its faculty of coverting sugars into starches; hence, a large part of its carbohydrate material remains in the form of sugar, although some starch may be developed.

Sweet corn culture is most extensive in the vicinity of large cities, where it is grown as a market-garden and truck crop, and in regions where it is grown as a canning crop.

188. According to the latest census, 1910, the number of farms reported as growing sweet corn in the United

States was 48,514, the number of acres, 178,224, the value of the product, $5,936,419. New York leads with the number of farms reporting, having 6,584, the number of acres being 23,739, and the value of the product $942,023. Pennsylvania is second in number of farms, 4,896 reporting sweet corn. The second place in number of acres, however, is with Illinois, 19,976 acres. Illinois is also second in the value of the product, having $558,746. Ohio is third in the number of farms, having 4,591. Maryland is third in the number of acres, reporting 18,387 acres devoted to the crop. In total value, New Jersey takes third place, with $557,708. Around the large cities of the northern part of the United States, large areas are devoted to the

FIG. 92. — An ear of green corn, at proper stage for table use.

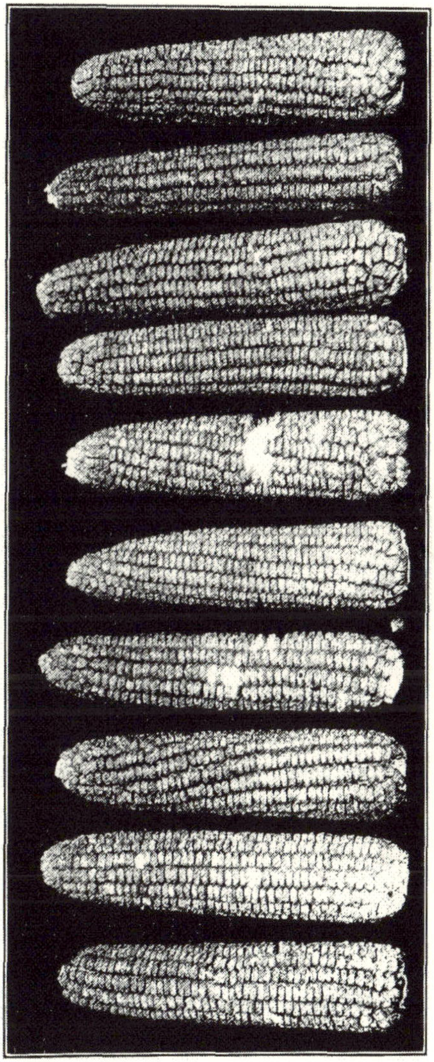

FIG. 93. — Sweet corn of the Evergreen type.

production of sweet corn for immediate consumption. Farther back and in several of the states, in particular Illinois, Indiana, Iowa, Maine, Maryland, New York, and Ohio, the great canning industry is developed, and large acreage is devoted to the growing of sweet corn for canning.

VARIETIES

The varieties can be classified under three heads: (1) canning, (2) commercial or market garden varieties, and (3) home garden varieties.

The principal canning variety is a late corn, Stowell Evergreen. Country Gentleman is also used to a large extent among the canners. In the Maine canneries, a corn known as Clark's is used. This is a corn which has been developed by the large canning houses, and the seed is grown in that section. Farther west, the Hickox is used to a considerable extent, while Trucker's Favorite and Evergreen are considered desirable in some sections.

189. The commercial varieties may be subdivided into three or four divisions. Among the early varieties are Cory, Adams Early, Early Maine, Peep o' Day, Aristocrat; medium earlies, Metropolitan, Golden Bantam, and other golden corns, Honey, Quincy Market, and Crosby; late varieties, Country Gentleman, Stowell Evergreen, Late Mammoth Hickox, Black Mexican.

In the home garden varieties, quality is of first importance. The custom is either to plant early and late corn, or one high-class variety such as Golden Bantam every two weeks for both early and late. A not unusual sequence of varieties is: Cory, Crosby, Quincy Market, Country Gentleman, and Stowell Evergreen.

SEED

190. The canning men as a rule raise their own seed, or have it raised on private farms by contract. In raising seed it is important to keep it free from contamination with other varieties, especially with field corn. When a sweet corn field is within a quarter of a mile of field corn, the sweet corn ear is likely to have kernels that resemble the field corn, due to wind-blown pollen. All blocks of seed corn should be far enough apart to protect against cross-pollination. The commercial grower or market-gardener very often produces his own seed.

In some cases the seed has been maintained on the same farm for many years. Some of these growers have succeeded by careful selection in developing desirable early types suited to their needs and as a result are able to market their product very early and secure the highest price.

191. The home gardener must depend practically from season to season upon the product that he can buy of the seedsman. If the seedsman is one who practices good methods of breeding and selecting his corn, the resultant seed is high class. If the seed is grown under contract and care is given, the results are satisfactory. When a large firm is responsible, it is reasonable to expect that the corn will come true to name.

192. Breeding and selecting sweet corn offers an interesting field for investigation. As explained before (page 105) the sweet corn grain type and starchy grain do not blend in hybridizing. The sweet corn type of grain is a recessive. This makes it possible to cross sweet corn with any type of starchy corn and, by selecting sweet corn grains from the hybrids, have pure sweet corns at once which may be combined with many or all the characters of the starchy parent.

SELECTING AND CURING SWEET CORN

193. Methods of selecting seed sweet corn vary with different growers. One way that has been found satisfactory is herewith given.

For many years it has been the custom to select for home planting a number of ears having characteristics

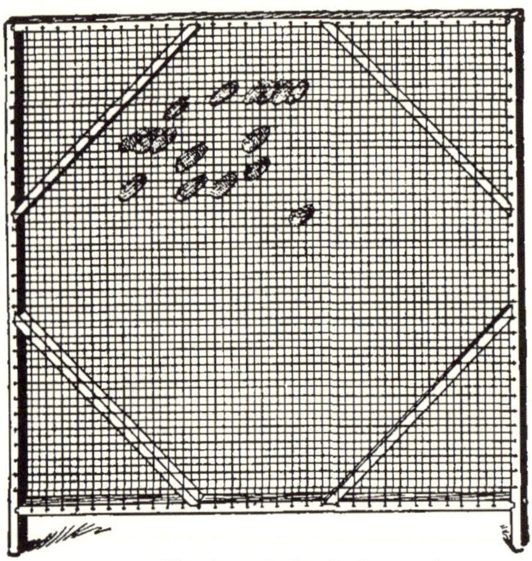

FIG. 94. — Handy rack for drying seed corn.

most desirable. Earliness is maintained only by saving the earliest ears from the early corn, and from these earliest ears a small number, known as "double extra," are set aside for the breeding-plat. In selecting these double extra ears, it is important to note, not only size, length of grain, and length of cob, but also the character of the corn for quality, as denoted by its translucent appearance. It is important to practice rigid selection, that is, not only to have a great many of the right kind of ears, but to plant none or the wrong kind in the breeding plat or near it.

One of the most important factors in the sweet corn industry is the proper curing of the seed ears. Sweet corn molds and ferments more easily than field corn. This greatly injures germination. Freezing before curing also injures germination.

194. Drying seed corn by fire heat is often practiced in seed houses equipped for the work, but is not the most practicable method on a small scale. Corn thrown in a large pile with or without the husk on will develop heat enough inside of twenty-four hours to injure the germ, sour the cob, and discolor the grain. Sweet corn cut and shocked up like field corn will sour before it dries, unless the weather be both cool and dry enough before winter to escape injury by freezing. Corn left on the stalk untouched until the husk opens will be greatly discolored and injured by a spell of hot, damp weather. If, however, the ears be husked out on a dry day and allowed to lie a few hours exposed to the direct rays of the sun, the organisms which cause fermentation are killed by the sunshine, and a layer of impervious matter is formed over the butt end of the cob, which makes it more difficult for fermentation to start.

The following method of curing sweet corn seed is recommended: When the husk is dead and loose on the ear, wait for a bright, clear day, begin early in the morning, and cut down a small piece of corn, throwing into piles. The same forenoon, when the sun is shining bright, husk it out as rapidly as possible, throw the corn into small piles on the ground, tie the fodder into bundles, and set it up in small shocks. Before night, haul in the corn and put it on a slatted floor. The floor is made of lath one inch thick by two inches wide, spaced one inch apart. The corn is taken up in baskets, and each basket is turned upside down on the slats, and taken off carefully, so that the ears are left like a pile of "jack-straws," crossed in every direction, many of them standing in a nearly vertical position. Each basketful of corn is emptied in a fresh place, and when all is done the slats are covered with corn about a foot deep, but so loosely arranged that there is no obstruction to the passage of air between the ears. In this position it dries very quickly and may be put into barrels as soon as all moisture is out of the cob. Each barrel may be covered with a piece of cloth held down by the top hoop, and then the barrel turned on its side.

This plan applies more to the regions with humid fall weather than to those regions in the West where fall weather is dry

GROWING SWEET CORN FOR CANNING

195. Canning corn is grown under contract with the firm in many corn-growing regions. The canning company sends out a contract similar to the following: —

Sweet corn agreement

(Place), (Date)

This agreement made with the —— Canning Company, by which I hereby agree to plant and raise for said Company —— acres of sweet corn, the same to be delivered at factory from time to time as required by said Company, in proper condition for canning during the season of 191–; for which said Company agrees to pay seven dollars per ton, said Company to furnish me at their factory seed corn at the proper time for planting. For said seed corn I agree to pay said Company two dollars per bushel on or before the first day of October, 191–, or from the proceeds of corn delivered on this contract. I further agree (1st) to plant said corn in three different plantings, first planting, —— acres, to be planted early in May; second and third plantings, —— acres, to be planted the last of May or the first week in June, or after each preceding planting is well up. (2d) Not to let any corn become heated or damaged by remaining in bulk too long, and to deliver said corn the day it is picked. (3d) To make a short snap close to the ear. (4th) It is further agreed that the corn covered in this contract shall not be paid for till October first, 191–. (5th) That corn must not be planted near field corn unless it be white field corn, as mixed yellow corn is unfit for canning. In case of destruction of the cannery by the elements, said Company not to be held liable for damages on this contract.

(Signed) (The Company)
(Signed) (The Farmer)

196. Rotation. — It is often to the advantage of the growers to plant their cannery corn crop in rotation with other crops. It is desirable that the corn can be planted following a sod, especially if on this sod from eight to ten tons of stable manure are applied and plowed under. From experiments, sweet corn is found to be greatly benefited by deep plowing in some soils. If choice of soil is obtainable, the piece of ground that will give the most satisfactory results is a gravelly or a sandy loam, especially if there is some chance of having humus, such as sod or manure. The corn is generally planted with a machine, either one- or two-row corn planter; and at the same time, some growers apply from three to five hundred pounds of a 3-8-5 fertilizer formula, or if the manure is deficient, up to 1000 to 1200 pounds of fertilizer of the same formula. In some cases, growers raising sweet corn place not only the above amount of manure on their ground, but sometimes more, and add the larger amount of fertilizer, as well.

197. Distance between the rows in planting is from 30 to 42 inches. Sometimes the distance between the hills in the rows is but 24 inches, and other times it will extend to 36 inches. The general custom is, with the smaller-growing varieties, to lessen the distance, whereas with the large-growing varieties, such as Stowell, the distance is increased so that each plant may have a normal amount of space for full development. More seed is generally planted in the hill than is required, from five to eight seeds being dropped in each. Later, this corn is thinned to three or four stalks to the hill. By this process the three or four best developed plants are allowed to remain.

The weeder is used soon after the seed is planted, or a fine-tooth harrow. When the corn has broken ground, the weeding is generally discontinued, and a fine-tooth cultivator used. This may be a one-row or a two-row cultivator. The general plan at first in cultivation is to till rather deeply, especially in the middle of the row between the plants, later tilling more shallow. The corn plant requires constant tillage and a good soil mulch for its best development and conservation of the moisture. Hand hoeing would be necessary if weeds were troublesome, especially if the plot was not check-rowed. However, there are some men that go to the extra care of check-rowing their corn, and cultivating in two directions, then omitting the hand hoeing. It may be an advantage to go through with a hand hoe, because, at the same time that hoeing is performed, sucker growths may be removed from the corn, thereby improving the quality and size of the ears. When it is seen that the horse and machine in cultivating are injuring the corn, this work is discontinued, and the corn is allowed to grow without farther attention.

About the time of marketing, the factories generally send a man to the field to instruct the farmer just when to bring the corn to the factory. In the different sections, there is some difference of opinion as to when the corn should be harvested for the factory and just how. In general, the corn should be delivered to the factory as soon as possible after breaking from the stalk. There are some companies that desire the corn broken in the morning and carted immediately to their factories. As stated in a number of reports received from canners, they did not desire the growers to pick the corn in the late afternoon and allow this to stand in the wagons over night, owing to heating of the corn.

In harvesting, the ear is broken from the plant so that there is very little or no stub left on the base, and the unnecessary husks as well are taken off. However, no extra attention or care is given at this period. The corn may be gathered in baskets or in boxes, and immediately emptied in a wagon. When the wagon is full, it is taken to the factory and there weighed, if sold at so much a ton green weight.

198. Thirty-five dollars is a fair return for an acre, exclusive of the value of the fodder, as well as the husk and the cob, which the growers can take back to their farms. The average returns for sweet corn to the acre are between three and four tons. Example: On good Iowa land, a farmer will average with a good stand about three tons per acre. Some years the yield will be as high as four and one-half tons. The price varies, but for large Evergreen corn from six to seven dollars is received per gross ton of corn with the husks on, and for smaller varieties the price is from $7.50 to $8.50 per gross ton. Other states report different yields and different prices for their corn. Very much depends on the cannery, the methods employed, and several other factors. The above are average figures.

Besides the corn grown for the canneries under contract, canneries often grow a large acreage of corn for their own use. The work there is conducted similarly to that of the men who contract with them. They plant their corn at different periods, so that it may extend over a long season and they may, by so planting, be able to keep the factory busy throughout the season.

MARKET SWEET CORN

Commercial corn growing for consumption in the green stage may be classed as: market-garden sweet corn growing, which embraces the extremely early and a small amount of the main season crop; and truck growing sweet corn, which never embraces the extremely early crop, but only the main and late crops.

199. The market-garden crop is generally grown on high-priced land near the centers of population. The soil is generally in the best condition and of the typical market-garden type, a sandy loam well supplied with humus, and improved each year by applications of manure, sometimes as high as 40 tons to the acre. Besides the heavy applications of manure, some market-gardeners use large quantities of commercial fertilizer. The general idea among them is that in order to get an early crop of sweet corn, which is the one that brings the highest money, they should have food for the plant quickly available.

200. From six to eight kernels, in some cases more, are planted in each hill. For the early varieties, the hills may be as close as one foot. From fifteen to eighteen inches is more nearly the average distance between hills in the row. The distance between rows varies from twenty-four to thirty inches. The cleanest culture is given, and irrigation is practiced in some cases.

Market-gardeners, by their intensive methods of planting, are able to place corn on the market from ten days to two weeks earlier than men living a little farther back from the centers of population, and practicing less intensive methods. In cultivating the corn, especially with the hoe, suckering is generally practiced.

Cultivation is continued thoroughly and as along as possible, the horse being muzzled when it is found that injury results. If the corn is not growing to suit, slight applications of fertilizer, especially nitrate of soda 100 to 150 pounds per acre, are made.

In planting the early and main season and late varieties, some planters practice sowing the seed at the same time, and allowing the difference in the period of maturity to bring the crop in at the proper time. Other growers prefer to plant their corn at intervals of ten days to two weeks. This latter seems to be the most practicable method.

201. Marketing. — As soon as the ear is at the right stage for harvesting it is broken from the plant and placed in baskets or boxes, immediately taken to the shed, and there repacked. In the eastern markets, especially in New England, the corn is packed in boxes, a certain definite number of ears in each box. For New York and Philadelphia and through the North and West, ears are sold by the hundred in sacks or hampers. This is less satisfactory. It is not a pleasing pack or one that attracts attention. The bushel box is more practical, more up to date and the corn carries better. In the sack the corn has been known to heat because too much was placed together.

202. The first corn coming to the market sells for thirty to forty and in some cases fifty cents a dozen. It then steadily declines until it reaches eight and even six cents a dozen. If a man has a retail route and has corn throughout the season, he usually maintains a high average price. Some men never sell for less than fifteen cents throughout the season from their retail wagons.

203. The bulk of the main crop and the late crop are grown a little farther back from cities on less expensive land, and under less intensive methods. The rows and hills are generally a little farther apart, three feet to forty-two inches between rows, and from thirty to thirty-six inches between hills in the row. Fertilizer up to a thousand or twelve hundred pounds is applied with the corn. The corn is commonly planted on sod ground, this being usually spring plowed. Clean culture is practiced in the early part of the season. The corn is generally harvested the same as for the market-gardening. When grading and packing is necessary, the ears should be of uniform size and about the same degree of maturity. Better prices can be thus secured. The corn is usually shipped to commission houses, to wholesale stores, to clubs and hotels. Gross returns of $100 an acre will make a crop of corn profitable. As high as $350 the acre has been received from sweet corn.

FORCING SWEET CORN

204. Forcing under glass has been practiced for commercial corn growing. Experiments have been tried, especially in New England. The Early Minnesota, Crosby, Early Cory, Adams and other varieties have been used for forcing with more or less success. A summary of suggestions is given here.

205. The requirements for forcing corn under glass are practically the same as those for forcing other warm-weather plants, such as tomatoes, melons, cucumbers, and egg-plants, — a day temperature of 70° to 80° and a night temperature of 60° to 70° being required, the atmosphere in the house to be rather moist during the first period of the corn's growth, but when pollen begins to fall, the atmosphere being dry. The crop should be marketed before July first, in order to be remunerative. Extra early varieties maturing in from 65 to 83 days from seed are to be used. The corn may be started in pots, either paper or clay, a few seed in each pot, and later transplanted where it is to stand in the greenhouse. Inter-cropping with radishes, lettuce, or spinach may be practiced, to utilize all space in the greenhouse to the best advantage. The distance between rows should be 18 inches, and between hills in the row 9 inches. Suckers are very common in a crop of this kind, and these should be removed. The principal pests in the greenhouse are rats and mice. They bother both by digging out the seed and by attacking the matured ear, spoiling it for sale. Poisoning or destroying these pests should be performed before the crop is planted.

206. Forcing corn in hot beds or cold frames very early in the season, allowing it to mature in these beds, is a practical method. In this way, corn may be obtained for consumption in the month of June when the pricc is very high. The general conditions of growth are the same as those for greenhouse work. The spacing between the corn is the same. Careful attention as to ventilation and watering should be given. The pollination in the hotbed or cold frame will be looked after by the natural elements, but in the greenhouse it is advisable to shake the corn plant slightly when the pollen is ripe.

A still later method of forcing has been practiced on a limited acreage near some cities, and that is starting the corn in paper pots or other receptacles. Two or three seeds are planted in each pot, allowing the corn to grow from four to six inches, and then transplanting the corn to the garden after the weather conditions have settled. The corn at this time should be four to six inches high. The roots have not suffered by being pruned, and the plant will continue its growth. This method has been tried both in the East and the Middle West, and where the demand warrants, has proved satisfactory.

SWEET CORN IN THE HOME GARDEN

207. In the home garden the aim should be to have a liberal and constant supply of sweet corn. The variety should correspond with the personal taste of the individual gardener or consumer. It is doubtful whether the extra early corns will answer the demands of the individual home gardeners, as they lack somewhat in quality.

The home gardener does not have a great choice of soil for the growing of sweet corn. The garden may be heavy clay or light loam. In either case the principal treatment should be liberal applications of stable manure. Some persons apply a little commercial fertilizer, but this is the exception rather than the rule. No fertilizer is needed if the garden has plenty of manure. Sweet corn in the home garden may be grown under the methods described for commercial growing. Transplanting corn from hotbeds is a feasible method for the home garden, especially for early corn. Inter-cropping of the corn, in the earliest stages when planted from seed, would be practical. Such crops as radishes, spinach, lettuce, and even beans can be grown in the home garden, utilizing apparently waste space, which later is necessary for the full development of the corn.

PART II

SORGHUMS

CHAPTER XXIII

THE SORGHUM PLANT

SORGHUM (*Andropogon Sorghum* var. *vulgaris*, Hackel, *A. Sorghum*, Brot., *Sorghum vulgare*, Pers.) is generally conceded to have been originally derived from the well-known wild species, *Andropogon halepensis*, Brot.

The wild species is found abundantly in all tropical and subtropical parts of the Old World and has been introduced into the Western Hemisphere, where it is now well distributed in both North and South America between the parallels of latitude thirty degrees north and south of the equator.

209. *Andropogon halepensis* is generally known in the United States as Johnson-grass. Johnson-grass is a coarse-growing perennial, with strong underground rootstocks by means of which it spreads rapidly and is very persistent, being regarded generally as a bad weed.

Sorghum differs from the wild form in that it is larger-growing, that it produces more seed, that certain forms have abundant sweet juice, and that no form is perennial or has persistent rootstocks. However, there are forms of *Andropogon halepensis* that are annual and without the persistent rootstocks, an example being the variety known as "Soudan grass." The wild form is somewhat variable, having certain types paralleling in their variations the cultivated forms.

GEOGRAPHICAL ORIGIN

210. Hackel[1] states that the cultivated forms had their origin in Africa, but Ball[2] believes that they also had an independent origin in India as well.

The early history of sorghum culture is unknown, but

FIG. 95. — Plant of sorghum. (After Fuchs, 1542.)

[1] HACKEL, EDWARD. The True Grasses, p. 59.

[2] BALL, CARLETON R. U. S. Dept. Agr., Bur. Plant Indus., Bul. 175, pp. 9–10.

there is good evidence that it was an important crop in both Africa and South Asia hundreds of years before the Christian Era. A reference to millet in the Bible (600 B.C.) probably refers to sorghum. (Ezek. x. 4. The word millet is translated "dochan" in the original Hebrew text, a word still in use in Arabic for various forms of sorghum.) Sorghum is well adapted to meet the needs of a primitive agriculture. The seeds provide human food, while the plant furnishes abundant fodder for animals. Under favorable conditions the plant will run wild to some extent, and is better able to care for itself than any other of our important cultivated plants.

Sorghum is at present the most important cereal food of the native people of Africa, and is a very important crop through the southern half of Asia. There are no statistics of the world's production of sorghum. The United States crop is estimated at about 3,000,000 acres and that of India at 25,000,000. The crop of Africa and of Asia Minor should approximate that of India.

BOTANICAL CLASSIFICATION

Order — *Gramineæ.*
Tribe — *Andropogoneæ.*
Genus — *Andropogon.*
Species — *A. Sorghum* var. *vulgare.*

211. Ball[1] has suggested the following classification as a key to the principal groups of sorghum: —

I. Pith juicy.
  A. Juice abundant and very sweet.
    1. Internodes elongated; sheaths scarcely overlapping; leaves 12–15 (except in Amber varieties); spikelets elliptic-oval to obovate, 2.5–3.5 mm. wide; seeds reddish brown. I. Sorgo

[1] BALL, CARLETON R. U.S. Dept. Agr., Bur. Plant Indus., Bul. 175, p. 8.

B. Juice scanty, slightly sweet to subacid.

1. Internodes short; sheaths strongly overlapping; leaves 12–15; peduncles erect; panicles cylindrical, spikelets obovate, 3–4 mm. wide; lemmas awnless. II. Kafir.
2. Internodes medium; sheaths scarcely overlapping; leaves 8–11; peduncles mostly inclined, often recurved; panicles ovate; spikelets broadly obovate, 4.5–6 mm. wide; lemmas awned. VII. Milo.

II. Pith dry.

A. Panicle lax, 2.5–7 dm. long; peduncles erect; spikelets elliptic-oval or obovate, 2.5–3.5 mm. wide; lemmas awned.

1. Panicle 4–7 dm. long; rachis less than one-fifth as long as the panicle.
   a. Panicle umbelliform, the branches greatly elongated, the tips drooping; seeds reddish, included. III. Broom-corn.
2. Panicle 2.5–4 dm. long; rachis more than two-thirds as long as the panicle.
   a. Panicle conical, the branches strongly drooping; glumes at maturity spreading and involute; seeds white, brown, or somewhat buff. IV. Shallu.
   b. Panicle oval or obovate, the branches spreading; glumes at maturity appressed, not involute; seeds white, brown, or reddish. V. Kowliang.

B. Panicle compact, 1–2.5 dm. long; peduncles erect or recurved; rhachis more than two-thirds as long as the panicle.

1. Spikelets elliptic-oval or obovate, 2.5–3.5 mm. wide; lemmas awned. V. Kowliang.
2. Spikelets broadly obovate, 4.5–6 mm. wide.
   a. Glumes gray or greenish, not wrinkled; densely pubescent; lemmas awned or awnless; seeds strongly flattened. VI. Durra.
   b. Glumes deep brown or black, transversely wrinkled; thinly pubescent; lemmas awned; seeds slightly flattened. VII. Milo.

212. Technical description. — The plant varies in height from about 4 feet (dwarf Milo) to 12 or 15 feet high in some of the tropical forms.

Panicle, or "head," varies in shape from the small, compact "sumac" type, in which the rachis is almost as long as the panicle, through the looser and more branching forms of the Collier type, in which the rachis is about one-half that of the panicle, to the broom-corn type, in which the rachis is only one-fifth the length of the branches.

Seeds. — The shape of seed varies, from round in the Kafir, Kowliang, and Shallu, to somewhat pear-shaped in certain of the sweet sorghums, somewhat flattened in Milo, and decidedly flat in the Durras. The seed coat of all dark-colored varieties has a decidedly astringent taste, due to the presence of tannin. The amount of tannin seems to vary with the color, being greatest in the black-seeded and dark red varieties, very little in yellow seeds, and there being none in white seeds. The astringency apparently has no ill effect except as it affects flavor, the dark-seeded grain not being so desirable for stock food on this account.

Stems. — Stems vary not only in height (from 4 to 15 feet), but also in relative thickness. The Amber variety is slender, with stems less than 1 inch in diameter, while in the Gooseneck variety the stems are 1 to 2 inches thick. In slender-stemmed varieties the nodes are usually long, about 12 inches; while in the stouter-stemmed varieties the tendency is toward short nodes, as in the Sumac, the average length being 8 or 9 inches.

Juices. — Stems are designated as juicy or dry. The actual water content of the green stems does not differ so much in the two cases, the green stems being 80 to 90 per cent water. In the juicy-stemmed varieties the juice is easily extracted by crushing and pressing. An ordinary roller cane press will extract 50 to 60 per cent of the juice.

Not all juicy sorghums are sweet, but practically all the very juicy varieties are. The sugar content of the juice in sweet sorghums varies from 10 to 18 per cent.

Leaves. — The leaves of the sorghums are strong and are especially well adapted to withstand the rather dry and often hot winds that prevail in semiarid regions. In periods of protracted drought the leaves assume a rather erect position, rolling together to a considerable degree in a way that appears to protect against excessive evaporation. All the very drought-resistant forms, as the Milo and Durra types, are rather scanty-leaved; the leaves being about eight to ten in number, rather broad and short, and rather coarse in texture.

Tillers. — All varieties of sorghum seem to produce tillers abundantly. These appear at the lower joints of the stem. The buds that develop into tillers may remain more or less dormant when conditions for growth are unfavorable, ready, however, to develop at the first favorable opportunity. Fertile soil and thin planting favor their development. Certain varieties, however, seem to produce two or more tillers normally, the tillers starting almost as soon as the main stem, and it is only under the very thickest planting that they are suppressed.

It sometimes occurs, when the first part of the season is dry and unfavorable, that the main stem may become stunted; if late rains come, the tillers will often grow much taller than the main stalk. The tillers are later in maturing and are considered undesirable when the crop is grown for grain or sirup; but they are usually desirable when the crop is grown for forage, as they no doubt increase the yield of fodder.

When sorghum plants are cut off, tillers usually spring up at once. In the South two crops, and even three crops, may be cut from the same roots. In regions of very mild winters the roots of certain varieties will live over, giving a crop the second year.

Branches. — Branches come from latent buds on the upper part of the stem as tillers do from the lower nodes. The same conditions that favor tillering favor the development of branches. The first branch appears from the *topmost* node, the second from the next, and so on down, in order; under very favorable conditions and thin planting, four or five branches may develop. Each branch bears a small head, similar to the main head but later in maturing.

Branches are considered undesirable, and the usual plan is to plant the sorghum thick enough so that there will be neither tillering nor branching.

Roots. — The Kansas station made a study of Kafir corn and sweet sorghum roots in comparison with corn and other field crops. The roots of Kafir corn were found to be finer and more fibrous than corn roots under the same conditions. A few of the longer Kafir roots penetrated to a depth of 3 feet, but most of them were confined to the upper 18 inches, filling the soil to this depth with a fine network of roots; while corn under the same conditions fully occupied the upper 30 inches with roots (see Fig. 12, page 27), sending its deepest roots about 4 feet. The sweet sorghum roots were somewhat intermediate in character, but resembled the Kafir more than the corn roots.[1]

The distribution of roots indicates that the sorghums draw their nutrients from the surface soil much more than corn.

[1] Kans. Agr. Exp. Sta., Bul. 127, pp. 207–208. 1904.

PHYSIOLOGY OF THE SORGHUMS

213. In general, the physiology and nutrition of sorghum are similar to those of corn, which has been set forth (page 38). The most interesting physical phenomenon of sorghum from an economic standpoint is its general resistance to drought and to the climatic conditions that prevail in dry climates.

Drought resistance. — The drought resistance of sorghum is well established. Its ability to yield in a dry climate is apparently not due to a deep root system or to any other adaptation of the root system so far reported. Neither does it seem to be due to a low water requirement, as the few tests made on this point indicate that quite as much is required per pound of dry weight as for Indian corn or for other crops not particularly adapted to dry conditions.

The success of sorghum under semiarid conditions seems to depend on two qualities, not found developed to so great a degree in other crops: (1) The high resistance of leaves to injury from hot, dry weather. The nonsaccharine groups, especially, will withstand dry and hot climatic conditions that would wither most vegetation beyond recovery. (2) The plants have the faculty of becoming almost dormant, so far as growth is concerned, for long periods during severe drought. During such periods the leaves roll and tend to assume an upright position. This, no doubt, reduces evaporation from the leaves and affords protection to the younger leaves and the seed head. The plant may remain in this condition, apparently without growth, for several weeks, far beyond the endurance of most cultivated plants. With the coming of rain, growth will usually be renewed with vigor. If the main stalk has been much stunted, tillers will often grow up at once and become taller than the main stalk. While tillers do not usually produce a good seed crop, they are satisfactory as forage.

REPRODUCTION

214. The sorghums are all "perfect-flowered" — the pollen and ovary being in the same flower, instead of in separate flowers as in corn. This is the principal botanical distinction between the tribe *Maydeæ*, to which corn belongs, and the tribe *Andropogoneæ*, to which sorghum belongs.

FERTILIZATION

215. All sorghums are adapted to both self-fertilization and wind fertilization. Apparently, self-fertilization is normal in the sorghums, and is in no way injurious as it is in corn (page 107). In developing pure strains of sorghum it has been found practicable to cover the heads with bags before blooming, thus securing complete self-fertilization.

NATURAL CROSSING

216. Under normal field conditions more or less crossing takes place. Regarding this point Ball[1] makes the following statement: "Just to what extent cross-fertilization takes place under normal field conditions, it is, of course, impossible to say. However, in the case of adjacent rows of different varieties, flowering on approximately the same dates, as high as 50 per cent of the seed produced on the leeward row has been found to be cross-fertilized. It is probable that in a fairly uniform field of any given variety a similar percentage of natural

[1] Ball, Carleton R. American Breeders' Association, Vol. VI, p. 193.

crossing takes place. Many writers have stated that such cross-pollination occurs also at very long distances, but this seems to be less conclusively proved. Probably a distance of 8 to 10 rods to leeward is the maximum at which appreciable hybridization occurs." Ball also states that the pollen is mostly shed during the early morning hours, when the winds are usually at lower velocities than later in the day.

Crossing of types. — All the different types of sorghum, as sweet sorghums, non-saccharine types, and broom-corns, cross readily. (See Fig. 115.) Broom-corn growers must exercise some care in keeping their seed stocks pure, in regions where other varieties of sorghum are grown.

CLIMATE AND SOILS

217. The entire botanical genus (*Andropogon*), made up of hundreds of species, is found growing principally in wide-open plains regions. Hackel[1] states, "the species prefer dry places, especially savannas."

Climatic requirements

Temperature and sunshine. — Sorghum, like corn, is a plant of tropical origin, varieties of which have been adapted to temperate climates. Like corn, it requires abundant sunshine and warm weather, being very sensitive to cool nights. At high elevations where nights are generally cool, sorghum seldom does well even when the days are warm and sunshiny.

Humidity and rainfall.—While both corn and sorghum require sunshine and warmth, they apparently differ somewhat as to humidity, corn preferring regions of high

[1] Hackel, Edward. The True Grasses, p. 57.

humidity such as prevail in the Mississippi valley, and sorghum preferring regions of dry air such as prevail in the Great Plains region of the upper Missouri River valley and southward.

The above general difference may be due in part to selection of varieties. Sorghum being of tropical origin and widely distributed, certain varieties flourish in very humid regions of Africa. Certain varieties of the sweet sorghums grow well in the Carolinas and Gulf States, where both rainfall and humidity are high.

While certain sorghums do well under humid conditions, the ability of all sorghums to remain more or less dormant during periods of drought, and to renew growth with the return of rain, has qualified the crop for adaptation to dry climates. For centuries sorghums have been grown and adapted to dry conditions in the Old World as they are being further adapted in the United States. The result is that the principal varieties of sorghum under cultivation prefer a drier and warmer climate than is required by the corn crop, although no doubt varieties of sorghum could be found equally adapted to humid regions. The above conclusion applies with more truth to the grain sorghums (Kafirs and Durras) than to the sweet sorghums or broomcorns.

Soil requirements

218. The sorghums are adapted to a wide range of soils, but they prefer a medium-weight loam to very light or very heavy soils. The grain sorghums are apparently more sensitive in this respect than the sweet sorghums. Sorghums for forage are often grown on poor land, not only because they produce more forage than any other crop under such conditions, but also because the stems are finer than when grown on heavy land.

219. Effect on the land. — The sweet sorghums sown thickly have the reputation of being "hard on the land." Grain sorghums planted thin seem to have the same effect also, in lesser degree. All millets have the same reputation. No very satisfactory explanation for this has been advanced. When the effect is noted it is most marked on the first crop following, and less marked afterward, usually completely disappearing in one or two years. The effect is most marked on small grain and less on intertilled crops.

As the sorghum roots are rather concentrated in the upper layers of soil, it is possible that this soil is very much exhausted of available fertility. There is some reason to believe that sorghums may exhaust available fertility to lower limits than do other crops. It is not known whether sorghums have a toxic effect on the soil.

The injurious effect when noted is considered only temporary, and farmers in general do not consider it a serious drawback to sorghum culture.

220. Alkali resistance. — Sorghum is often said to be alkali-resistant. It is not resistant in the same sense as are many native alkali plants, but at least it is one of the best of our cultivated plants to succeed on land rich in alkali.

SORGHUM TYPES

221. A common grouping, based principally on the economic use of the crop, is (*a*) Saccharine sorghums, (*b*) Non-saccharine sorghums, (*c*) Broom-corns.

A. Saccharine sorghums. Those having an abundant sweet juice. Cultivated at one time principally for sirup manufacture, but now principally as a forage plant. Commonly known as "sorghum." I. Sorgo.

B. Non-saccharine sorghums.

1. Pith contains a scant juice, which varies from slightly sweet in some varieties to subacid in others. Grown principally for the grain, but also has forage value. II. Kafir.
III. Milo.

2. Pith dry

(*a*) Grown principally for the grain and forage.
II. Kafir.
VI. Durra.
IV. Shallu.
V. Kowliang.

(*b*) Grown for the bush, no value as forage.
VII. Broom-corn.

The economic discussion of sorghums will follow the above grouping.

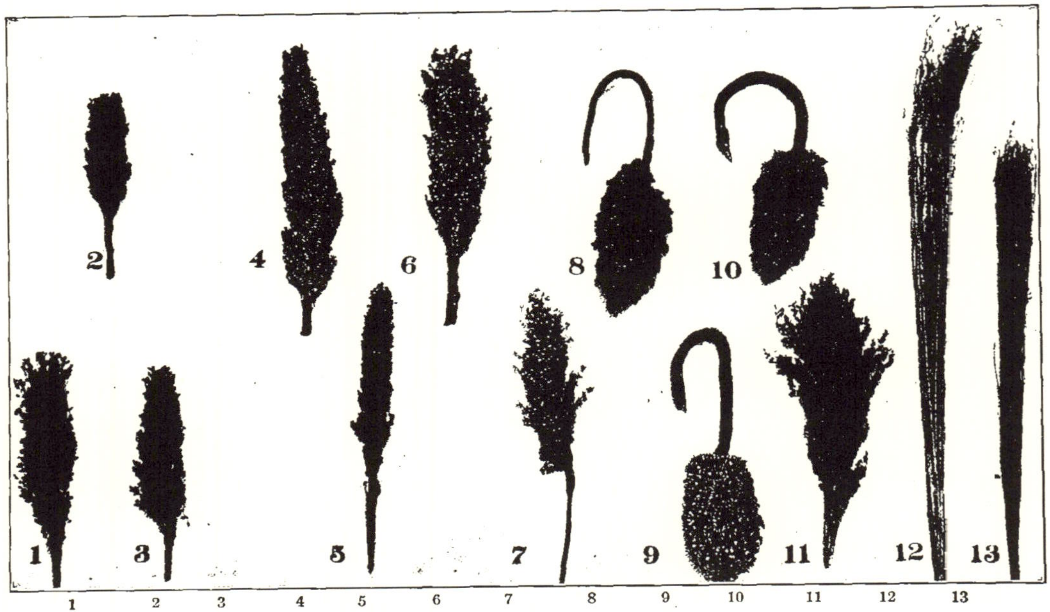

FIG. 96. — The principal types of sorghum.

1. Amber. 2. Orange. 3. Sumac. 4. Red Kafir. 5. Pink Kafir. 6. Blackhull Kafir. 7. Shallu. 8. Milo. 9. White Durra. 10. Brown Durra. 11. Brown Kowliang. 12. Standard broom-corn. 13. Dwarf broom-corn.

CHAPTER XXIV

THE SACCHARINE SORGHUMS

Sweet Sorghums

222. This group of sorghums is usually designated as sweet sorghums, or "sugar" sorghums. They are quite distinct from the non-saccharine, *grain sorghums* in having a juicy stem containing a high percentage of sugar and in producing a rather light seed crop.

Early culture. — The sweet sorghums have never been cultivated extensively in the Old World, where the sorghums have been cultivated more for seed than for forage — the non-saccharine forms being more productive for the former purpose. The sweet sorghums seem to have been kept in cultivation principally for the sweet canes, which, however, were not manufactured but were peeled and the juice was expressed by chewing. Almost no sweet sorghum is raised in North Africa or in India; it has been kept in cultivation in China and South Africa, however, though only in a small way.

223. Introduction into the United States. — The first recorded introduction into the United States was from China in 1853, by way of France, and the plant was known at first as "Chinese Sorgo." This was a loose-panicled sorghum, from which have been derived most of our cultivated varieties of Amber sorghum. "Our Early Amber is said to have originated in 1859 as a sport in a field of Chinese sorgo growing in Indiana." [1]

[1] Ball, Carleton R. *l.c.*, p. 25.

A collection of sixteen varieties of sorghum brought from Natal, South Africa, to Europe in 1854 and from Europe to this country in 1857, included several sweet sorghums, from which have been derived our compact-headed types such as Orange, Sumac, and Gooseneck.

Development of culture in the United States. — While sweet sorghum has remained a secondary crop in the Old World, it had a rather rapid development in the United States, owing to the belief that it would become a great sugar- and sirup-producing crop. In 1857 the United States Patent Office distributed 275 bushels in small lots to farmers; *The American Agriculturist* distributed to its subscribers 1600 pounds in small packages, and the next year 34,500 pounds in the same way. At this time extensive experiments were being made with it in Europe for the manufacture of sugar, and later the United States Government[1] conducted an elaborate series of experiments for the same purpose. With the development of sugar beets at this time a better source of crystallized sugar was found, and the plan of using sorghum for this purpose was abandoned.

First grown as a sirup crop. — However, sorghum was found to be a cheap source of home made sirup and it was more or less grown for this purpose in every rural community. Local " sorghum mills " were very common during the eighties in the Central and Western States. During the dry years in the early eighties, and again during the general drought of 1892–1894 in Nebraska, Kansas, and southward, sorghums of all kinds were found to withstand drought.

There are no available data on acreage of sweet sorghum, but the data on Kafir corn (page 304) indicate the

[1] See U. S. Dept. Agr., Bur. Chem., Buls. 26, 40, etc.

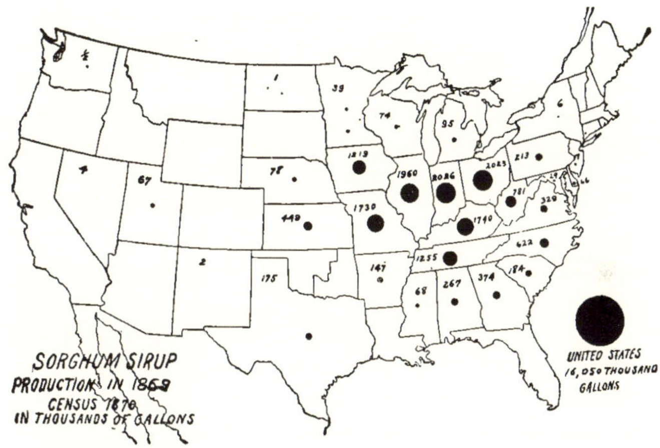

FIG. 97. — Production of sorghum sirup middle of last century.

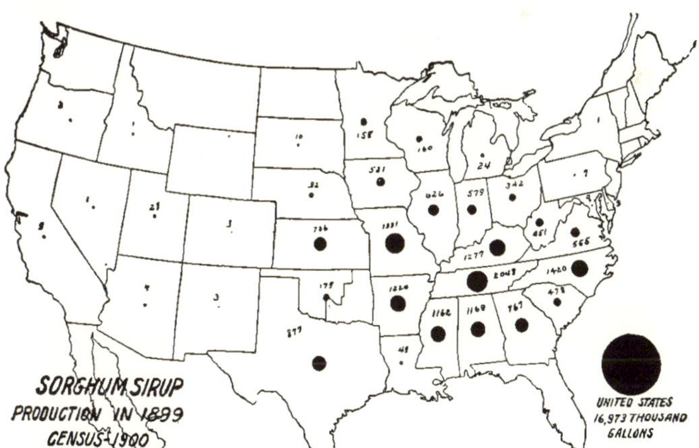

FIG. 98. — Production of sirup at close of century.

increase of sweet sorghum as the acreage of the two crops of late years is about equal. Beginning with 1890, the acreage has continued to increase up to the present time.

224. How the crop is utilized. — In the Central States east of the Mississippi River, these sorghums have been cultivated principally since 1865 for the manufacture of sirup. The extent of sirup manufacture for the census years is as follows: —

Year	Gallons
1860	6,749,123
1870	16,050,089
1880	28,444,202
1890	424,235,219
1900	16,972,783

The principal States in sirup manufacture for the last three decades have been Tennessee, Missouri, and Kentucky, but the industry has shown a rapid decrease in all these States. In only one State, North Carolina, has it shown a notable increase. Figures 97 and 98 show graphically the distribution at two periods.

225. As a forage crop. — West of the Missouri River and southward in the Great Plains region, the culture of sweet sorghum is principally as a forage crop. It is an important forage crop in the drier parts of Kansas, Oklahoma, Nebraska, and Texas. The use of sweet sorghum as a forage crop has developed since 1880.

226. Classification of sweet sorghums. — The following classification is adapted from Ball: —

A. Peduncle and panicle erect.

1. Panicle loose, open, branches spreading to horizontal or drooping; rachis two-thirds as long to equaling the panicle.

Empty glumes black, hairy. I. Amber.
Empty glumes black, smooth. II. Minn. Amber.
Empty glumes red. III. Red Amber.
Empty glumes light brown. IV. Honey.

Rachis less than one-half the length of the panicle:—
Panicle light, drooping branches, seeds orange to red. V. Collier.
Panicle heavy, seeds orange. VI. Planter's Friend.

2. Panicle close, compact.
Empty glumes equal to seeds, seed red. VII. Orange.
Empty glumes half as long as the small seeds, seeds dark red. VIII. Sumac.
Empty glumes narrow. IX. Sapling.

B. Peduncle recurved (goosenecked) or sometimes erect.
Panicle black, glumes awned. X. Gooseneck.

The three varieties that have had most extensive cultivation are Amber, Orange, and Sumac.

227. Amber, being the earliest of the three (90 to 100 days), has been practically the only variety grown in the northern limits of sorghum culture — that is, north of Kansas and the Ohio River — and has been most popular in Kansas, the leading sorghum-growing State.

Amber grows about 5 to 7 feet tall, with 8 to 10 leaves, being neither so tall nor so leafy as the other two varieties. The seed head is usually black and is loose or spreading, though it is somewhat variable in this respect. A number of selections have been made, the best known of which are: *Minnesota Amber*, which differs only in minor details; *Red Amber*, the heads of which are red instead of black but which is otherwise similar; and *Folger's Early*, a strain said to be especially desirable for sirup production. The various strains of Amber sorghum have been popular for forage because of the rather slender stems and early maturity, these qualities facilitating the curing and improving the quality of forage.

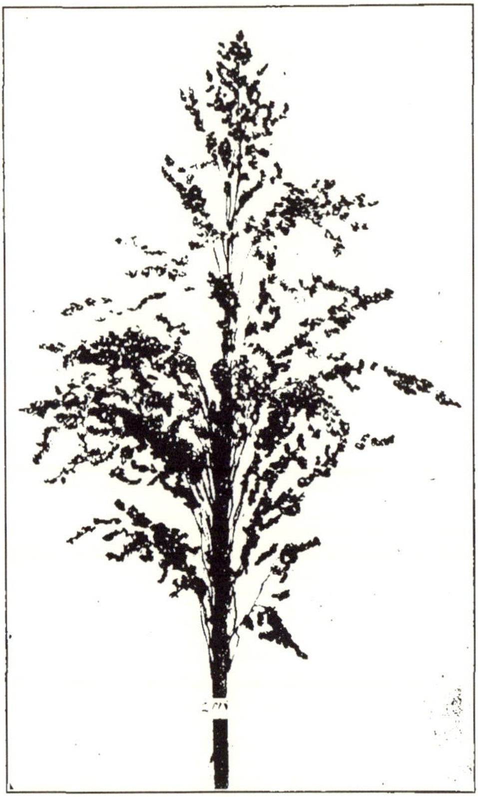

Fig. 99. — Amber sorghum.

228. Orange sorghum is two to three weeks later in maturing (100 to 125 days) than is Amber. It is about 12 inches taller, the stalk is heavier and the nodes are shorter, and the plant is more leafy. The variety name refers to the deep orange color of the ripe heads. This variety is excellent for sirup production and it makes a heavy yield of forage, especially on good land. However, for cured forage farmers object somewhat to heavy stalks, as they are more difficult to handle and cure. Orange sorghum is second in popularity to Amber and is grown principally from Kansas southward.

FIG. 100. — Orange sorghum.

Collier and Coleman are two varieties of the Orange sorghum type which are so similar to it that for all forage purposes they may be considered the same. The Collier is considered the better for sirup-making.

229. Sumac sorghum derives its name from the very compact red seed head, resembling the seed head of sumac. It is somewhat larger and perhaps later than Orange, but otherwise similar in appearance of plant. "For forty years this has been the most popular variety in the South, especially in the Piedmont districts. It is now largely grown in Texas and Oklahoma also."

230. Gooseneck is a very large, late-growing variety, adapted only to the South. Ten to fifty per cent of the heads are recurved, or "goosenecked."

Fig. 101. — Sumac sorghum.

Fig. 102. — Gooseneck sorghum.

CHAPTER XXV

THE NON-SACCHARINE SORGHUMS

231. The non-saccharine sorghums, with the exception of broom-corn, are often called *grain sorghums* because their principal value is as grain producers rather than as producers of forage. As a group, they constitute the most drought-resistant grain and forage crops in cultivation. The five principal types of the non-saccharine sorghums are: (1) Kafir, (2) Durra, (3) Shallu, (4) Kowliang, (5) Broom-corn.

Historical. — The non-saccharine sorghums are very generally cultivated throughout Africa, southwest Asia, India, and Manchuria, but are not cultivated extensively in Europe. In general, the kafir types dominate in South Africa, the Durra types in North Africa, southwest Asia, and India, and the Kowliang types in Manchuria. Shallu, the least important of the five principal groups, is grown as a winter crop in India, and the same type has been reported as grown in a limited way in Madagascar and at several points in Africa.

232. The Durra group (spelled also dura, durah, doura, dhoura, and other ways) is the most important in the Old World. It should be noted, however, that there are three general groups of the durra sorghums, only one of which is important in the United States: (1) The types grown in central and northeast Africa are tall, large-seeded, and late-maturing, furnishing both forage and grain; (2) those

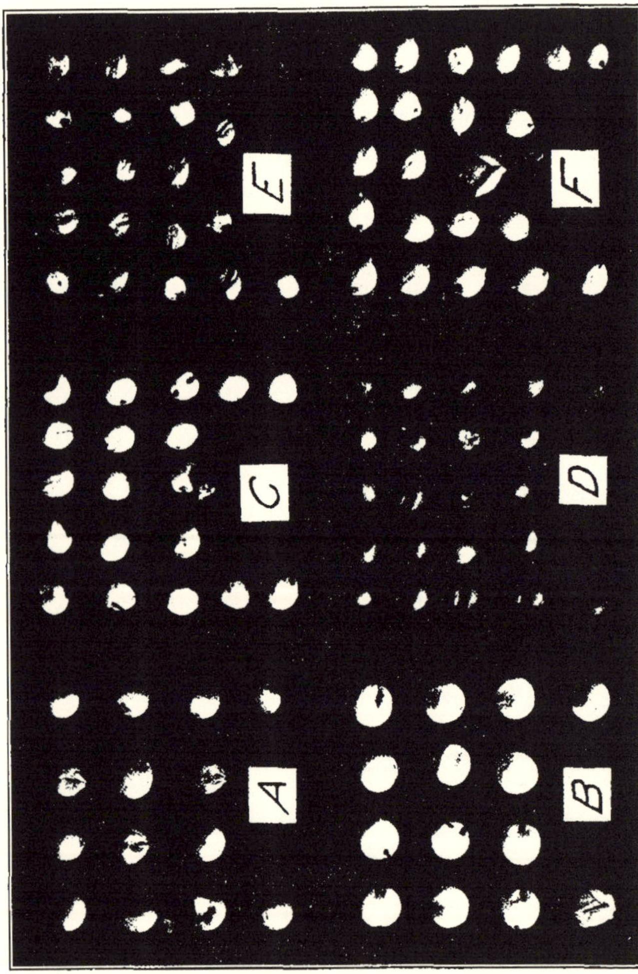

FIG. 103. — Seeds of grain sorghum: *A*, Milo; *B*, White Durra; *C*, Blackhull Kafir; *D*, Red Kafir; *E*, Brown Kowliang; *F*, Shallu. (Slightly enlarged.)

of North Africa are shorter, early, comparatively low in forage and high in grain production, and the grain is flat and of medium size; (3) those of India have comparatively small heads and seeds, the seeds not decidedly flat; they produce both forage and grain, but are too large and late-maturing for culture in the United States.

The second group has thus far furnished most of the varieties that have found a place in United States agriculture. The probable reason is that grain sorghum could not compete with maize in the corn-growing belt. There was, however, a distinct demand for crops adapted to the Great Plains, a region too dry for the culture of corn. The sorghums from the more humid regions of the Old World have not always been drought-resistant, and in most cases are too late in maturing. Most of the kafirs and durras meeting the requirements of drought resistance and a short maturing season have come from the drier regions of North Africa and the high plains of South Africa.

233. Introduction in the United States. — The cultivation of non-saccharine sorghums dates from the introduction of White Durra and Brown Durra into California in 1874 and the introduction of kafir in 1876, but they were not generally distributed until about ten years later.

234. Region where cultivated. — The "grain sorghums" are cultivated for grain and for forage. They are not so desirable for forage alone as are the sweet sorghums; the fodder is coarser and lacks the sweet sugars in the stem, being less palatable. They are commonly harvested for both grain and forage. As a grain crop they cannot compete with corn in the regular corn-growing belt, and therefore the principal grain-sorghum belt lies just west of the corn-growing belt, following in general the line of 25-inch rainfall on the east and extending west to the Rocky

Mountains; the belt includes also southern California and Utah. The accompanying chart[1] (Fig. 104), prepared

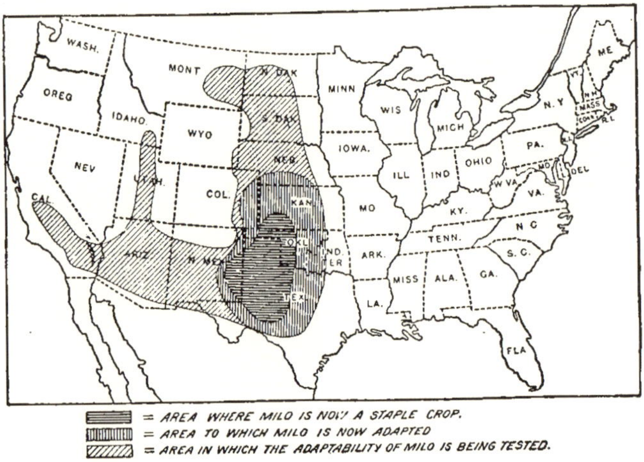

Fig. 104. — This map made to show the distribution of milo; also shows, approximately, the area where the culture of all sorghums are of most importance.

to show the area of Milo culture, outlines the probable area of grain-sorghum culture.

235. Statistics of culture. — It is not possible to supply exact figures on the production of grain sorghums. The census of 1909 gives the total acreage of kafir grown for grain as 266,513. The principal States reported, and their acreage, were as follows: —

State	Acreage
Kansas	154,706
Oklahoma	63,455
Texas	22,813
California	20,218
Total	261,192

[1] U. S. Dept. Agr., Farmers' Bul. 322, p. 11.

Ninety-eight per cent of the entire acreage was produced in four States. The above figures do not include that sown as forage. From the Kansas State Board of Agriculture, however, we have data for the census year showing 631,040 acres sown for forage, or about four times as

FIG. 105. — Two heads of Milo, showing good and poor types.

much as that harvested for grain. On the same basis for the United States it would appear the the non-saccharine-sorghum acreage for 1909 was about 1,250,000 acres. The acreage has since increased slightly in Kansas and to a marked degree in Oklahoma and Texas, so that present acreage is above two million acres. The value of non-saccharine sorghums is now recognized, and with the im-

Acreage, Value, and Yield of Kafir, Milo, and Corn for the Years 1904 to 1909, inclusive, in Kansas and Oklahoma

KANSAS					
Crop and Year	Acreage	Yield per Acre in Tons or Bushels	Value		
			Per Ton or Bushel	Total	Per Acre
Kafir:					
1904 . . .	518,372	3.04	$3.19	$5,041,546	$9.70
1905 . . .	538,393	3.24	3.06	5,352,810	9.91
1906 . . .	548,497	3.05	3.01	5,039,238	9.18
1907 . . .	508,485	2.94	3.78	5,658,860	11.11
1908 . . .	630,096	2.85	3.82	6,856,845	10.89
1909 . . .	636,201	2.79	4.02	7,150,080	11.21
Average .		2.99	3.48		10.33
Milo:					
1904 . . .	7,166	3.18	3.22	73,476	10.24
1905 . . .	20,550	2.84	3.28	190,974	9.31
1906 . . .	17,563	2.55	3.26	146,289	8.31
1907 . . .	22,090	2.72	3.90	234,686	10.61
1908 . . .	55,255	1.92	4.85	515,269	9.31
1909 . . .	102,492	1.97	4.74	959,259	9.34
Average .		2.53	3.87		9.52
Corn:					
1904 . . .	6,494,158	20.3	.39	50,713,955	7.81
1905 . . .	6,799,755	28.0	.36	68,718,584	10.11
1906 . . .	6,584,535	28.4	.35	65,115,203	9.25
1907 . . .	6,809,012	21.3	.43	63,040,743	9.26
1908 . . .	7,057,535	21.3	.55	82,642,462	11.71
1909 . . .	7,711,879	19.1	.57	83,066,905	10.77
Average .		23.1	.44		9.83

Acreage, Value, and Yield of Kafir, Milo, and Corn for the Years 1904 to 1909, inclusive in Kansas, and Oklahoma

OKLAHOMA

Crop and Year	Acreage	Yield per Acre in Tons or Bushels	Value		
			Per Ton or Bushel	Total	Per Acre
Kafir:					
1904 . . .	334,948	9.79	$0.40	$1,312,204	$3.92
1905 . . .	297,286	12.72	.40	1,512,318	5.08
1906 . . .	269,218	16.10	.34	1,465,937	5.44
1907 . . .	371,405	13.50	.58	2,881,032	7.77
1908 . . .	400,047	9.20	.46	2,548,200 [1]	6.36
1909 . . .					
Average		12.26	.44		5.71
Milo:					
1904					
1905 . . .	138,608	20.06	.40	1,112,602	8.02
1906 . . .	122,347	17.82	.40	870,767	7.12
1907 . . .	131,366	13.30	.65	1,142,098	8.64
1908 . . .	145,096	12.55	.33	757,565 [2]	5.22
1909 . . .					
Average .		15.93	.45		7.25
Corn:					
1904 . . .	1,369,276	16.00	.39	8,544,339	6.24
1905 . . .	1,369,276	16.00	.39	8,544,339	6.24
1906 . . .	1,642,930	18.90	.40	12,436,557	7.56
1907 . . .	1,528,735	31.40	.36	17,142,081	11.21
1908 . . .	4,014,631	18.10	.48	35,409,961	8.82
1909 . . .	4,284,561	18.60	.48	38,449,866	8.97
Average .		20.60	.42		8.56

[1] Includes $828,131 worth of fodder.

[2] Includes $151,911 worth of fodder.

provement of varieties they are destined to become an important crop west of the 98th meridian. The comparative acreage and value of non-saccharine sorghums compared with corn in Kansas and Oklahoma, as compiled in Bulletin 203, Bureau of Plant Industry, United States Department of Agriculture, is given above.

236. Classification of non-saccharine sorghums. —

KAFIR GROUP

Pith juicy:
(Very juicy, sweet = *Sorgo.*)
Juice scanty, subacid or somewhat sweet or dry in certain varieties.

(1) Heads erect, cylindrical, spikelets oval, small, 3–4 mm. wide.

(*a*) Seeds white:
- Glumes greenish white, some darker. — I. White Kafir.
- Glumes black or nearly. — II. Blackhull Kafir.

(*b*) Seeds red:
- Glumes deep red to black. — III. Red Kafir.

DURRA GROUP

(2) Heads pendent but sometime secret, ovate; spikelets broadly obovate, large, 4, 5–6 mm. wide

(*a*) Seeds white:
- Glumes greenish white, silky, seeds flattened, awned. — IV. White Durra.
- Glumes black, seeds smaller, less flattened, rare. — V. Blackhull Durra.

(*b*) Seeds yellowish to reddish brown:
- Glumes short, wrinkled, reddish to black, not silky; seeds yellowish brown; florets awned. — VI. Yellow Milo.
- Glumes as long as seeds, greenish white, seeds reddish brown, not awned. — VII. Brown Durra.

Broom-corn Type

Pith dry:
Head loose, 10–28 inches long; spikelets oval or obovate, small, 2.5–3.5 mm. wide, lemmas awned:
Rachis one-fifth as long as branches.
(*a*) Branches drooping, seeds reddish. XI. Broom-corn. Rachis more than two-thirds as long as head:
(*b*) Branches of panicle drooping; glumes at maturity spreading and involute; seeds white to buff (several varieties). VIII. Shallu.
(*c*) Branches spreading but not drooping, glumes at maturity appressed, not involute; seeds white, brown, or red. (Several varieties, corresponding to the red, white, and blackhull varieties of Kafir and Durra. Also standard and dwarf.) IX. Kowliang.
Head compact, erect or pendent, spikelets oval or obovate, small, lemmas awned:
Rachis two-thirds as long as head. X. Kowliang.

237. Kafir. — The three principal varieties of kafir are red, white, and blackhull. The heads are erect, in contrast to the durra group, in which the heads are mostly recurved, or "goosenecked." The white and blackhull varieties both grow about 5 to 6 feet high, while the red is 8 to 12 inches taller. The white and red varieties were first introduced. The white variety, however, was not satisfactory because of its not maturing well, and the head was not always exerted from the leaf sheath, thus inducing rot in damp weather. The red variety matured properly and soon became more popular.

The objection to Red Kafir was the astringent taste of the seed coat, common to all kafirs with a colored seed coat. The blackhull, a white-seeded variety, appears to be a later introduction, having attracted attention about 1896. It had all the good qualities of Red Kafir, and in addition the seed was not astringent. This variety probably furnishes nine-tenths of the kafir crop to-day, and Red Kafir the other tenth.

FIG. 106. — Plant of Blackhull Kafir.

238. Durra. — The characteristics of this group are that the heads are mostly "goosenecked" and the seeds are large and flat. The extensive culture of non-saccharine sorghums in this country began with the introduction of Brown Durra and White Durra into California in 1874, but the culture did not become general in the Great Plains region until about 1890.

The White Durra is commonly known as "Jerusalem corn," and sometimes as "Egyptian rice corn." The Brown Durra is often called "Egyptian corn."

White Durra is little grown, as it is frequently injured by insects and diseases. The grain also shatters badly.

Brown Durra has continued in cultivation especially in southern California and Texas. The total area of White and Brown Durras was estimated at 50,000 to 60,000 acres in 1908.[1]

239. Milo, or Yellow Milo, was introduced about 1885, ten years later than the White and Brown Durras, but it quickly became the most popular of the group, the area in 1908 being estimated at 300,000 acres. This variety will mature in 90 to 100 days and is adapted to culture as far north as southwestern Nebraska. In addition to the standard varieties, there is now a dwarf variety well suited to cultivation for grain production.

Compared with kafir, the durras are better adapted as grain producers but not so well suited for forage production. Milo is the best suited of all the sorghums for grain production. Early varieties of milo have been developed by selection, which adapts it to a

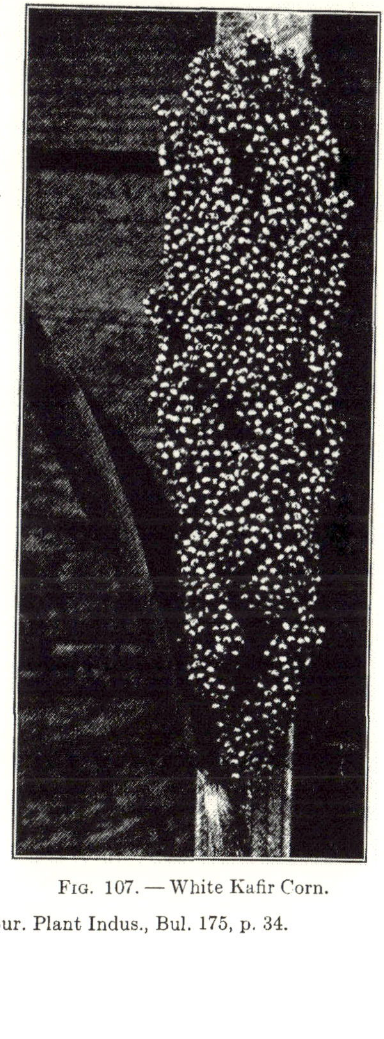

FIG. 107. — White Kafir Corn.

[1] U. S. Dept. Agr., Bur. Plant Indus., Bul. 175, p. 34.

FIG. 108. — Upright Milo head.

FIG. 109. — Pendent form of Milo head.

wide range of conditions, and this plant, together with Blackhull Kafir, is the best of the sorghums for grain production.

The milos, being about three weeks earlier in maturing than the kafirs, have two distinct advantages: in Oklahoma and Texas they can be planted early and will more nearly mature before the severe midsummer drought; also, they may be grown farther north and at higher altitudes.

Fig. 110. — Yellow Milo.

240. Shallu. — This plant is of recent introduction. The stalks are tall and slender, with large loose and open panicles, approaching broom-corn in type. The plant comes from India, where it is cultivated as a winter crop, being sown in October and harvested in March. It is grown for both seed and forage. Seed of this was introduced and tested by the Louisiana Agricultural Experiment Station about twenty years ago. It is occasionally grown from Kansas to Texas. It has acquired several local names, as California wheat, Egyptian wheat, and Mexican wheat.

241. Kowliang. — In both India and China the sorghums are commonly classed with millets. " Kowliang," or " tall millet," is a Chinese name given to distinguish this variety from the common smaller millets (Panicum and Chætochloa). The three colors of seed and glume found in kafirs and durras are found also in this group, namely, brown seeds with black glumes; white seeds with black glumes, and white seeds with white glumes. There are varieties of both dwarf and standard size, 4 to 11 feet high.

Kowliang comes from northeast China and the adjacent territory of Manchuria, 38° to 42° north latitude — the farthest north of any region where sorghums have been an important crop for any great length of time. They are extensively cultivated in this region for grain and forage and the stems are used for fuel.

All varieties are early-maturing, and, being already adapted to a region farther north than any other group of sorghums except the Early Amber varieties (the original Amber type also came from China), they should be adapted to a similar latitude in the United States. They have not been extensively tried in this country, but the early dwarf stocks give promise of furnishing a good foundation stock for the development of grain sorghums in the northern half of the Great Plains. They could not replace Early Amber sorghums as a forage crop.

CHAPTER XXVI

CULTURAL METHODS FOR SORGHUMS

242. Sorghums are grown for four distinct purposes: (*a*) as a grain crop primarily, (*b*) as a forage crop, (*c*) for sirup manufacture, and (*d*) for broom-corn brush.

The land to be chosen would be similar in each case, but the principal difference in cultural methods would come in method of sowing and harvesting.

Because the sorghums will grow on poorer and drier land than any other of our cereals is to be taken as an indication not that they naturally prefer such conditions, but rather that they are capable of withstanding greater hardships than other crops. Consequently, the culture of sorghums may extend beyond the limits of common cereals; but, on the other hand, they will respond as readily to manuring and to favorable environment as will any plant, on good, rich land producing six to seven tons of cured forage per acre.

Preparation. — The land is prepared much as for corn. The plowing may be done in the fall or in the spring. As planting does not take place until rather late — two to four weeks after corn, — there is ample time for spring preparation of the soil.

GROWING SORGHUMS FOR GRAIN

243. Varieties. — Blackhull Kafir, Milo, Red Kafir, and Brown Durra, in the order named, are the principal sorghums grown for grain.

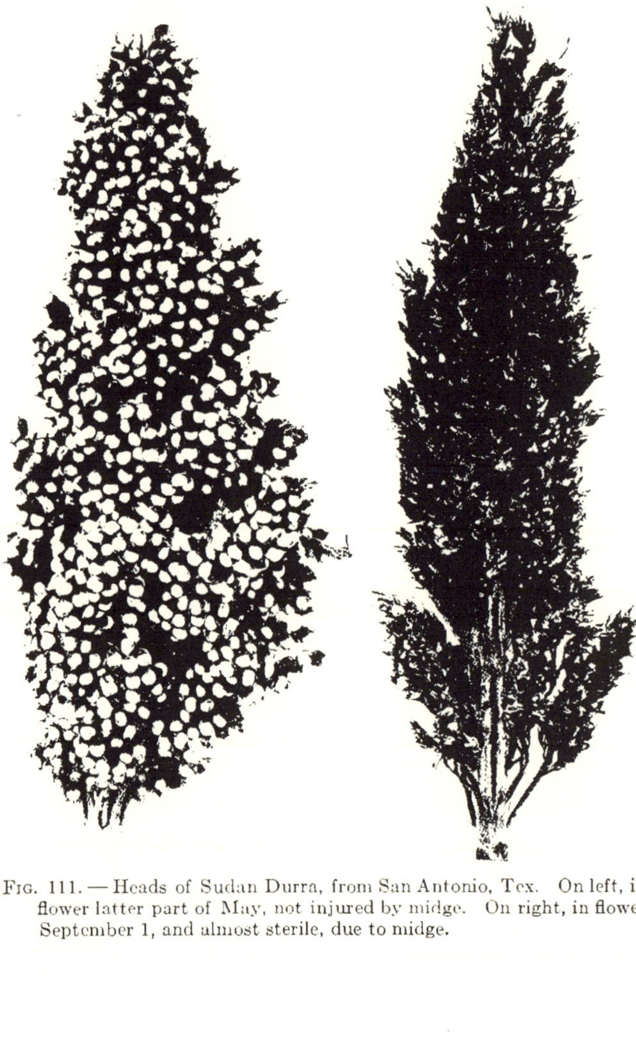

FIG. 111. — Heads of Sudan Durra, from San Antonio, Tex. On left, in flower latter part of May, not injured by midge. On right, in flower September 1, and almost sterile, due to midge.

244. Time of Planting. — Grain sorghums are usually planted soon after corn; the time ranging from March to June in the Southern States, while as far north as Nebraska the planting must be as early as possible in order to insure maturing. Planting in Nebraska practically coincides with corn planting, about May 10.

In the San Antonio region of Texas it has been found necessary to plant very early in order to avoid the sorghum midge, an insect that becomes very numerous in June and practically prevents all seeding from that date on. In order to avoid the midge, planting must be early. According to one experiment reported in 1911, eleven varieties of grain sorghums planted on March 4 yielded 23.1 bushels, while early varieties planted on March 15 gave only profitable yields, and no varieties planted on April 1 were profitable.[1]

245. Rate of planting. — Grain sorghums are usually planted in rows 3 or $3\frac{1}{2}$ feet apart; the plants 6 to 8 inches apart for the milos and durras, and 8 to 10 inches for kafirs. On very fertile soils the planting should be thicker than this. The amount of seed required will be 3 to 5 pounds per acre. With durras a higher percentage of the heads "gooseneck," or recurve, when planted thin than when planted thick.

246. Methods of planting. — Corn-planting machinery is generally used for sorghums, the only change necessary being to use special plates for dropping or to adapt the corn-dropping plates. The corn-planting plates can be adapted by filling the holes with lead and boring out to the right size. Grain sorghums are always drilled.

Listing is a method common in regions of low rainfall,

[1] Grain Sorghum Production in the San Antonio Region of Texas. U. S. Dept. Agr., Bur. Plant Indus., Bul. 237. 1912.

but in regions of higher rainfall and heavy soils surface planting is better. When planted in a lister furrow the seed should not be covered deeper than is necessary to

FIG. 112. — Plat of Milo selected for erect heads.

insure good germination, as it rots very easily when planted deep or when the soil is cold or wet.

Surface planting is ordinarily done with the two-row corn-planter; the grain drill is sometimes employed, however, by using only every fourth or fifth hole.

247. Tillage. — The same tools are used in general for cultivating sorghum as for corn, and in much the same manner. However, sorghum, especially when listed, is much slower in growth than corn for the first four weeks, and consequently more skill is required to clean out the weeds. Young sorghum is tougher and less likely to break than is young corn, which is an advantage, since it permits of the use of such tools as harrows and weeders oftener and longer than is the case with corn. With surface-planted sorghums, by the proper use of harrows and weeder it is often not necessary to give more than one thorough cultivation with the shovel cultivator.

With listed sorgum, the harrow and lister cultivators should be used for the first cultivation. When the plants

FIG. 113. — Field of White Kafir in shock.

are 8 to 10 inches high a very thorough cultivation should be made with the cultivator, to be followed later by such shallow cultivation as is necessary to keep down weeds.

248. Cutting. — When grown for grain the heads should be fully ripe. If cut for silage, the seeds should be in the soft dough stage, as the ripe seeds in silage are very likely to pass through the animal without digestion.

The corn-binder is the best and most economical implement for harvesting on a large scale. With smaller areas the sled cutter is used, or the crop is cut by hand.

Various plans for harvesting only the heads have been tried, but all these have proved less satisfactory than harvesting the whole plant.

249. Curing. — The grain sorghum, however harvested, should be set up in shocks until well cured. Precaution should be taken to set the base of the shock wide and to tie well about the heads. The heads being heavy, the shocks are very likely to fall over.

Before threshing, the sorghum heads should be very dry, as the grain heats and spoils quickly when stored if at all damp. This will require four to six weeks in the shock.

250. Hauling and storing. — Where the fodder is fed, it is very common to haul from the field as used. Sorghum will remain in very good condition for several months when bound and set in large shocks. If not to be used for three months, it is usually better to haul and stack.

Baling is sometimes practiced, a hop or broom-corn baler being used as the bundles are not broken apart.

When the stover and grain are to be fed separately the bundles are sometimes beheaded with a broadax or heavy knife. The heads are then stored in a dry place, to be fed whole or to be threshed.

251. Threshing. — The whole bundles are sometimes run through an ordinary grain-thresher, or only the heads run in and the bundles then withdrawn. The labor is heavy in both cases and it is often considered better to behead the bundles and thresh only the heads.

Yields

252. As shown by the table on page 306, the average yield of grain sorghums in Kansas and Oklahoma is not equal to that of Indian corn; but in these States corn is raised in the part of the State having heaviest rainfall, and sorghum in the drier part.

West of the 25-inch-rainfall line, grain sorghums will equal or outyield corn. The advantage increases as rain decreases. Yields of twelve to twenty bushels of grain sorghum are often harvested when corn is a failure from drought. Twenty bushels per acre is considered an average crop and forty bushels per acre a good crop. Yields of seventy bushels have been known.

GROWING SORGHUMS FOR FORAGE

253. Sweet sorghums are used more extensively when grown primarily for forage than are the non-saccharine.

Since the foliage of all sorghums remains green until the heads are mature, a fair quality of coarse forage is secured when sorghums are grown for grain. About one-half the sorghum crop is sown primarily for fodder, to be cut before heads are ripe and cured as fodder or hay.

254. Time of planting. — In the Gulf States sorghum is often sown early so that the crop may be cut two or three times, though sowing may continue for several months. In the Central States sowing is usually after corn planting, generally in the month of June.

255. Rate of planting. — Sorghum for forage is either sown thick in drill rows about 3 feet apart and cultivated, or sown close, either broadcast or with the grain drill.

When sown in rows to be cultivated, the methods are similar to those for growing grain except that about 15 pounds of seed per acre is used instead of 2 to 5 pounds.

When sown broadcast, one to two bushels per acre of seed are used; the thinner sowing is done on poorer land or in a dry climate, and the thicker seeding under the most favorable conditions.

256. Methods of planting. — Which of the two methods shall be employed — drilling or broadcasting — depends on circumstances. In regions of low rainfall, drilling in wide rows and cultivating is the surer method, but in more humid regions there is little difference in yield. On the other hand, drilling in rows increases the cost because of the amount of cultivation necessary. The fodder is also coarser.

Harvesting forage sorghum

257. When cultivated in rows the best method of harvesting is with a corn-binder. The bundles are set up in small shocks to cure. In four to six weeks several small shocks may be set together in large shocks, which

FIG. 114. — Cutting sorghum forage with a mower.

are securely tied near the top and left in the field to be hauled as used. A better method is to stack in large stacks, but care must be observed that the fodder is well cured before stacking.

When sown broadcast the crop is usually cut with a mower and handled as coarse hay, or cut with the grain-binder

When cut with a mower a stubble of 6 inches should be left. This tall stubble facilitates drying, and also gathering the heavy fodder with a hayrake. Heavy sorghum hay dries very slowly and should be left for one to two weeks in the swath before raking and cocking. It should be thoroughly cured in the cocks before stacking.

258. An average yield of cured fodder varies from 3 to 6 tons per acre. Very heavy yields of 10 tons per acre have been reported from one cutting. Where sorghum is cut two or three times a season, as in the South, the relative yield of the different cuttings depends on the method of handling. If the first cutting is allowed to become quite ripe, the following cutting will be light; but if the first crop is cut quite green, the second cutting may be as heavy as, or heavier than, the first.

259. Seed crop. — Twenty-five to thirty bushels of seed per acre is considered an average yield. All sorghum sown in rows for fodder or planted thin for sirup-making produces a good crop of seed. Most of the commercial seed of sweet sorghums comes from this source.

CHAPTER XXVII

UTILIZING THE SORGHUM CROP

260. In Asia and Africa the grain of sorghum is utilized principally as human food, in the United States as stock food.

The seed coat is hard and rather indigestible, therefore all sorghum grain fed to live stock should be ground.

Composition. — The composition of kafir is shown by the following summary:[1] —

FOOD CONSTITUENTS IN KAFIR. IN FRESH OR AIR-DRY MATERIAL

	WATER Per Cent	ASH Per Cent	PRO-TEIN Per Cent	FIBER Per Cent	NITRO-GEN-FREE EXTRACT Per Cent	FAT Per Cent	AUTHORITY
Kafir (whole plant green) .	76.13	1.75	3.22	6.16	11.96	0.78	Penn. Station
Kafir (whole plant green) .	76.05	1.44	2.34	8.36	11.41	0.40	N. Y. (Cornell) Station
Average .	76.09	1.60	2.78	7.26	11.69	0.59	
Kafir fodder (whole plant .	10.94	5.48	3.31	30.37	47.40	2.50	N.C. Station
Kafir fodder (without heads) . . .	8.67	7.14	4.89	28.02	49.75	1.53	Kans. Station
Kafir (mature head) . .	16.23	2.02	6.92	6.79	65.18	2.86	N.C. Station
Kafir seed .	9.31	1.53	9.92	1.35	74.92	2.97	Kans. Station
Kafir flower .	16.75	2.18	6.62	1.16	69.47	3.82	N.C. Station

[1] Cycl. of Agr. *IV*: 387.

Kafir and other sorghum seeds are considered to be very starchy foods. For good results they require that some protein food, as alfalfa hay or cottonseed meal, be fed with them. Ten per cent cottonseed meal is sufficient. Kafir grain fed alone is also constipating, and this tendency is corrected by the addition of a protein food fed in connection.

When fed to cattle, horses, and sheep, good results are secured, though pound-for-pound feeding experiments show sorghum to be not quite so valuable as corn. In general, for fat stock, 80 to 90 pounds of corn have been found to equal 100 pounds of kafir or milo when fed in comparison.

261. Poultry food. — Sorghum seed is one of the best poultry foods and enters into a large proportion of these foods found on the market. It is considered superior to corn. For poultry the seed need not be ground but is fed whole, either threshed or in the head.

262. Soiling or green feed. — Sorghum is probably the most popular crop to cut and feed green. The sweet sorghums are used principally for this purpose. The superiority of sorghum for this use lies in its large yield, its sprouting up from the roots so that the crop may be cut several times in succession, and its drought resistance. Sorghum will remain green and growing under drier conditions than will other forage crops, furnishing succulent food at the time it is most needed.

For green feeding it is usually drilled very thick, in rows 3 feet apart.

An acre of green sorghum producing 12 tons will feed twenty head of stock for twenty days, allowing 60 pounds per head each day.

263. Pasture. — Sorghum is used considerably as a

pasture crop. For this purpose it is sown rather thick, 2 to 3 bushels per acre. Stock is turned in when the crop is 3 to 4 feet high.

For pasturing, the field should be divided into lots and enough stock should be turned in to eat down the crop in about two weeks. The stock should then be removed to another lot and the pasture given four to six weeks to grow up again. This would require three to four lots.

It is estimated that one acre will furnish grazing for the equivalent of one animal for one hundred days, or ten animals for ten days.

264. Sorghum mixtures for pasture. — For pasture purposes German millet is sometimes mixed with sorghum and gives good results. Cereals have been used as a mixture, but it is doubtful whether they add to the value as pasture. In the South, it has been recommended to mix sorghum and cowpeas, for both forage and pasture. Cowpeas give a better-balanced ration. For pasture the sorghum and cowpeas should be drilled in rows about 8 to 12 inches apart, in alternating rows.

265. Sorghum for silage. — Within the corn-belt, sorghum compares favorably with corn as a silage crop. In regions of less than 25 inches rainfall, sorghum will probably come to be the most important silage crop. In the South, also, it is likely to supersede corn for silage, especially where the crop is to be grown on rather poor land.

Sorghum silage is more difficult to preserve than corn, being more likely to ferment. When well preserved it appears to have a feeding value about equal to that of corn silage, though very little experimental work on this point has been done. Sorghum for silage is now in extensive use in many places in the Southern States.

266. Sorghum poisoning. — Sorghum pasture under some conditions is a virulent poison. This is due to prussic acid forming in the leaves under certain conditions. The conditions favoring the development of prussic acid seem to be hot, clear, and dry weather, producing a stunted growth. Poisoning is most common in semiarid regions. When conditions are right for developing poison, the sorghum should be pastured with caution, as the poison acts quickly and there is no known remedy. Cattle should not be pastured on stunted or drought-stricken sorghum. Where it is desired to test the pasture, probably the best way is to allow only a single animal to graze the field for a day or two.

When poisonous sorghum is cut and allowed to lie until wilted, the poisonous property entirely disappears.

CHAPTER XXVIII

SORGHUM FOR SIRUP-MAKING

As discussed heretofore (see page 296), sorghum has had an extensive use in the United States for sirup manufacture. The process of sirup-making is so simple that nothing more is necessary than a roller press for extracting the juice, and a single evaporating pan. In a few cases rather extensive plants have been established, but most of the sirup has been made in small local plants.

267. For sirup the sweet sorghums are used, as Amber, Orange, Sumac, and Gooseneck. There are strains of all these varieties selected for sirup-making. (See description of these varieties, pages 297–300.)

268. For sirup the sorghum is planted and cultivated practically as described for the culture of grain sorghums.

269. Time of harvesting. — The sugar content of sorghum at different stages of growth as determined by Collier, the result of 2740 analyses, is given as follows:[1] —

SUGAR CONTENT OF SORGHUM AT DIFFERENT STAGES OF GROWTH

STAGE OF CUTTING	SUCROSE	INVERT SUGAR
	Per Cent	Per Cent
Panicles just appearing . .	1.76	4.29
Panicles entirely out . . .	3.51	4.50
Flowers all out	5.13	4.15
Seed in milk	7.38	3.86
Doughy, becoming dry . .	8.95	3.19
Dry, easily split	10.66	2.35
Hard	11.69	1.81

[1] Sorghum Sirup Manufacture. U. S. Dept. Agr., Farmers' Bul. *477*: 12.

270. Sorghum increases not only in total weight until mature, but also in the percentage of sugar. The seed should reach a hard dough stage before cutting.

Stripping. — For best results the leaves should be stripped. This is done while the canes are standing. The canes are often pressed without removing the leaves, but if this is the case, the yield of juice is less and the impurities are much greater.

Cutting. — The canes are cut by hand or with a corn-binder. In hot weather, cutting should be done not more than two days before grinding, as there is danger of fermentation developing. In cool fall weather, however, canes are often kept in large shocks for one to two weeks after cutting.

When a heavy frost occurs the sorghum should be cut and placed in large shocks at once. If it is to stand for some time, both leaves and heads should be left on. In large shocks, with cool weather the sorghum may be kept with little loss for three or four weeks.

A heavy freeze will do no harm provided the cane can be ground at once upon thawing; but after thawing it is likely to go out of condition in a very short time.

271. An average yield of green sorghum would be 8 to 10 tons, though it may vary from 5 to 15 tons.

The yield of sirup depends on the kind of mill, quality of the sorghum, and quality of the juice.

A poor mill may extract only 30 per cent of the total juice, while with a good three-roller mill 60 per cent of the original weight may be extracted as juice, or 1200 pounds to a ton of canes.

Juice varies in quality, containing 8 to 15 per cent of sugar. The juice is concentrated by boiling until it contains about 70 per cent of solid matter and 30 per cent of water. The amount of sirup produced from a ton of canes is therefore very variable.

In general, a ton of canes will give 700 to 1200 pounds of juice, which in turn will yield 10 to 30 gallons of sirup, according to quality.

272. The manufacture of sorghum sirup consists of three steps: (1) extraction of juice; (2) clarification of the raw juice; (3) evaporation of juice.

The extraction is done with heavy roller presses of either the two-roller or three-roller type. The juice is then run into settling tanks, where impurities in suspension are allowed to settle out.

The clarification is accomplished in some cases by merely allowing the raw juice to settle for some time. Settling is hastened by heating. Sometimes fine yellow clay is added, which aids in settling. When the juice is somewhat acid, lime also is added to the heated juice. After clarification the clear juice is drawn off to be concentrated.

Concentration takes place in large, shallow pans, where the juice is kept boiling by a well-regulated fire. Ordinarily the pan is divided into compartments, the boiling juice flowing slowly in a thin layer from one end to the other. By the time the outflow is reached, the juice should be concentrated into sirup. In very small plants the juice is merely boiled down in kettles.

CHAPTER XXIX

BROOM-CORN

Broom-corn belongs to the non-saccharine sorghums, resembling Shallu or Kowliang more than others. It is characterized by very short rachis and long, slender, seed-bearing branches. The plant is grown principally for the seed head, or " brush," having practically no forage value.

273. Historical. — The origin of broom-corn is not known, though it was cultivated and used for making brooms two hundred and fifty years ago[1] in Italy, where it apparently had its first general culture. References are made to its culture in the United States about the year 1800.. The following statement appears regarding it in a book entitled " The Pennsylvania Farmer," published in 1804:[2] " A useful plant, the cheapest and best for making brooms, velvet whisks, etc. The grain for poultry, etc., a few hills or rows of it in the garden or cornfield suffice for family purposes."

While its value was thus recognized, its culture did not become important until several decades later.

274. Statistics of culture. — During the past forty years, broom-corn culture has developed rapidly, as shown by the crop harvested for the past three census years: —

Year	Pounds
1879	29,480,106
1889	38,557,429
1899	90,947,370

[1] Mentioned by Casper Bauhin as used for this purpose in 1658.

[2] Twelfth Census. Vol. VI, Part II, p. 519.

The crop practically trebled in thirty years.

Broom-corn culture has always been concentrated to certain rather limited regions: Four States in 1879 — Illinois, Kansas, Missouri, and New York — produced 80 per cent of the crop. In 1889 four States, the first three named above and Nebraska, produced 89 per cent of the crop. In 1899 the last-named four States and Oklahoma produced 90 per cent of the crop.

In 1899 Illinois alone, which has been the leading State in broom-corn production for forty years, produced 66.7 per cent of the entire crop in the United States, while 50.1 per cent of the entire crop was grown in three counties.

The twenty-two counties of the United States producing more than 1000 acres each are shown in the following table, as reported by the Twelfth Census: —

FIG. 115. — Broom-corn, sorghum, and hybrid between the two: *a*, broom-corn; *b*, hybrid; *c*, black-seeded sorghum.

County	State	Acres	Pounds Produced	Average Yield per Acre
Coles	Illinois	34,597	23,948,030	692
Douglas . . .	Illinois	22,356	14,768,780	661
Moultrie . . .	Illinois	10,256	6,815,530	665
Cumberland .	Illinois	6,619	2,738,710	414
Edgar . . .	Illinois	6,248	4,085,860	654
Woods . . .	Oklahoma	6,086	1,292,670	212
McPherson . .	Kansas	5,684	2,890,330	509
Reno	Kansas	5,137	1,691,090	329
Rice	Kansas	4,167	1,366,030	328
Henry	Missouri	3,753	1,177,950	314
Shelby . . .	Illinois	3,246	1,826,670	563
Clark	Illinois	2,446	1,210,140	495
Henry . . .	Illinois	2,000	1,298,450	649
Allen	Kansas	1,952	566,480	290
Cass	Nebraska	1,726	776,580	450
Stafford . . .	Kansas	1,684	553,710	329
Jasper . . .	Illinois	1,496	651,560	436
Piatt	Illinois	1,454	950,710	654
Sheridan . .	Kansas	1,307	305,910	234
Cheyenne . .	Kansas	1,090	252,940	232
Stevens . . .	Kansas	1,054	267,680	254
Polk	Nebraska	1,051	498,000	474

275. Varieties. — Seedsmen list broom-corn under at least a dozen variety names, but these names have little significance. There are two types, known as (1) standard, normally growing about 12 feet high with a brush 18 to 28 inches in length, and (2) dwarf broom-corn, growing 4 to 6 feet in height and producing a brush 12 to 18 inches in length.

The standard type is used for the manufacture of large brooms.

While dwarf brush is also used to some extent in the manufacture of large brooms, the straw is generally too fine and weak for this purpose. The dwarf type, however, is almost exclusively used in whisk brooms. There is some variation in different strains. Very often the large manufacturers keep on hand seed of the strains best suited to the needs of the trade, and are ready to supply growers with this seed.

276. Brush. — The brush should be bright and of a uniform light green color. When the head does not fully exsert from the " boot," or upper leaf sheath, the base of the brush is likely to take on a red color, which is very undesirable. The discoloring is most common when considerable rain occurs during the maturing season. This is a very common fault of the dwarf variety and necessitates breaking over the brush as soon as it is well grown so that it will hang down. For this reason dwarf broom-corn is more successfully grown in rather dry climates, most of it at present being cultivated in Kansas and Oklahoma.

Length of brush. — In general, the longer the brush the better, all other qualities being equal. There is some danger that very long brush may be coarse. Brush that is both fine and long is the most valuable.

Rachis. — The rachis should be short, with no central " core " of stiff branches extending upward in the center.

Shape of head. — The head should be broom-shaped rather than conical, with all branches approximately the same length.

Flexibility. — The brush should be flexible and tough. This condition is attained both by proper climatic conditions and by proper harvesting.

277. Culture of broom-corn. — The selection and preparation of land, method of planting, cultivating, and so on, are no different in general from those in the culture of other sorghum crops. However, quality and uniformity in the crop is as important as yield, and more precaution must therefore be taken to have the land uniform, and the

FIG. 116. — Poor and good heads of standard and dwarf broom-corn (after C. P. Hartley): *a*, poor head of dwarf with large center; *b*, head of dwarf inclosed in "boot"; *c*, good grade of dwarf for whisks; *d*, long head of dwarf with characteristic weakness at point *x*; *e* and *f*, good grades of standard hurl; *g*, good head of self-working; *h*, poor grade of standard because of heavy center; *i*, smutted head.

stand uniform. Also, the cost of harvesting is much increased if the crop does not ripen so that it can all be harvested at one time.

Land. — Any productive soil will raise broom-corn. The principal consideration is that the soil be uniform. One reason why the culture of this plant has been so successful in central Illinois is because of the extensive areas of uniform soil.

Planting

278. Time of planting. — The planting of broom-corn usually begins about two weeks later than the planting of field corn and may be continued for a period of four weeks. In the Central States, planting is done from the middle of May to the end of June and harvesting begins the middle of August. It is often desirable to distribute the planting so that the harvesting will not come too much at one time.

Method of planting. — The width of row varies from 3 feet for dwarf varieties to $3\frac{1}{2}$ feet for standard varieties. The distance apart in row is 2 inches in dwarf and 3 inches in standard varieties. The planting should be uniform, as the brush will be too coarse where the stalks are thin, and undersized where the planting is too thick.

Drilling is the ordinary method of planting. The ordinary corn-planter, with special plates for broom-corn seed, is satisfactory.

Replanting thin places is not practicable, and thinning the stand is too expensive. It is, therefore, very important to take every precaution to secure a perfect stand at the beginning. It is hardly necessary to state that the land should be clean and in good tilth, and the seed should be carefully cleaned and of good germinating quality.

279. Tillage. — The same tools and methods of cultivation that are successful with Indian corn are effective with broom-corn, except for the fact that broom-corn is more delicate and grows slowly the first three weeks, necessitating greater care and skill.

280. Time of harvesting. — In order to get a good green color and tough, flexible brush, the corn must be cut quite green, or just as soon as the brush has reached full growth. The best time is when just past full bloom.

If allowed to ripen, the brush loses color and becomes brittle, and the selling price for such brush is often less

FIG. 117. — Standard broom-corn, tabled and ready for hauling.

than one-half that of high-grade stock. On the other hand, when allowed to ripen, 10 to 20 bushels of seed per acre is secured, which is valuable as a poultry and stock food. It is generally conceded that the loss in value to the brush is much greater than the value of the seed crop, although in California the seed crop is quite generally harvested; but this is not customary in other places.

Cutting the brush. — Dwarf broom-corn is usually "pulled," while the standard type is "tabled" and cut.

Dwarf varieties are short enough so that a man can easily reach the heads; also, the base of the brush is inclosed in the "boot," which must be removed. When the crop is uniform enough so that all can be pulled at one time, the cheapest way is to pull and load directly on wagons. When it must be pulled twice, that harvested the first time over is laid on the ground and covered with leaves. It is not possible to get a uniform grade in this way.

Standard broom-corn is first "tabled" and the heads are then cut by hand. In tabling, one man passes backward

FIG. 118. — Threshing broom-corn seed heads or brush.

between two rows, bending the stalks at a point about 30 inches above the ground toward each other and across the row, so that the heads hang about two feet past the other row. Two men following cut off the heads and place them evenly, on every other table. Three men can harvest about two acres per day. Later, a team with a wagon passes over the empty tables and the brush is collected.

Threshing and storing. — The heads are threshed directly from the field, or within a very few days after cutting. The thresher removes all seeds, after which the brush is stored in drying sheds, in thin layers about 3 inches deep.

Bulking. — After drying for about three weeks the brush is piled in tiers, called " bulking," for further drying. It

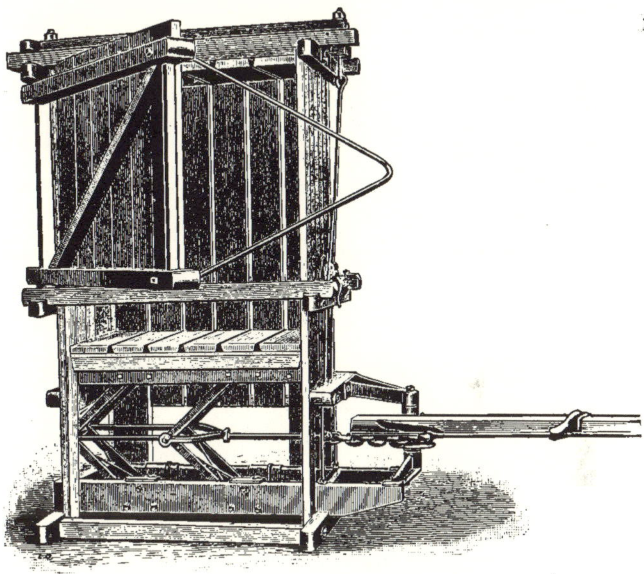

FIG. 119. — Power baling press for broom-corn.

then goes " through the sweat," which means merely that considerable natural heat is developed and the drying is hastened.

Baling. — This should not take place until the brush is thoroughly dried. Good bales of brush are often very much damaged by heating and molding, as a result of baling before dry. A bale weights 300 to 400 pounds.

281. Market grades. — Certain trade terms are applied in describing the qualities of broom-corn, which are well

Fig. 120. — A bale of broom-corn.

understood by those familiar with the stock. The following data, prepared by C. P. Hartley, give trade terms and relative prices of different grades: —

	Cents per Pound
Fair, crooked	$1\frac{1}{2}$
Good, well-handled, crooked	2
Fair, medium, red-tipped	$3\frac{1}{2}$
Slightly tipp d, smooth growth	4
Good, green mooth, self-working	$4\frac{1}{2}$
Choice, green, self-working carpet stock	5
Fair, medium, sound hurl	$3\frac{1}{2}$
Good medium hurl	4
Good, green, smooth, carpet hurl	5
Choice, green, smooth, carpet hurl	$5\frac{1}{2}$

REFERENCES ON SORGHUMS

Bureau of Plant Industry, United States Department of Agriculture: —

Bulletin 50. Three Much Misrepresented Sorghums.

Bulletin 175. The History and Distribution of Sorghums.

Bulletin 203. The Importance and Improvement of Grain Sorghums.

Bulletin 237. Grain Sorghum Production in the San Antonio Region of Texas.

Farmers' Bulletins, United States Department of Agriculture: —

Bulletin 37. Kafir Corn Characteristics and Uses.

Bulletin 50. Sorghum as a Forage Crop.

Bulletin 92. Improvement of Sorghum.

Bulletin 174. Broom Corn.

Bulletin 246. Saccharine Sorghums for Forage.

Bulletin 288. Non-saccharine Sorghums.

Bulletin 322. Milo as a Dry Land Crop.

Bulletin 334. Sorghum for Silage.

Bulletin 448. Better Grain Sorghum Crops.

Bulletin 450. The Best Two Sweet Sorghums for Forage.

Bulletin 477. Sorghum Sirup Manufacture.

Kansas Agricultural Experiment Station Bulletins: —

Bulletin 23. Smuts of Sorghum, Corn Smut.

Bulletin 93. Kafir Corn.

Bulletin 99. (Page 5.) Kafir Corn, Alfalfa Hay, and Soy Beans for Pork.

Bulletin 99. (Page 32.) Kafir Corn.

Bulletin 99. (Page 35.) Digestion Experiments with Kafir Corn.

Bulletin 119. (Page 28.) Kafir Corn *vs.* Good Butter.

Bulletin 119. (Page 42.) Sorghum Pasture for Dairy Cows.

Bulletin 119. (Page 63.) Whole Kafir Corn Compared with Ground Kafir Corn for Growing Calves.

Bulletin 136. (Page 164.) Sorghum with Corn for Baby Beef.

Bulletin 136. (Page 179.) Kafir Corn Meal and Sorghum Seed Meal with Soy Bean Meal for Swine.

Bulletin 136. (Page 202.) Kafir Corn with Alfalfa for Baby Beef.

Bulletin 149. Prevention of Sorghum and Kafir Corn Smut.
Oklahoma Agricultural Experiment Station Bulletins: —
Bulletin 22. Field Experiments with Kafir Corn.
Bulletin 35. Summary of Digestion Experiments with Kafir.
Other Bulletins: —
Florida Agricultural Experiment Station, Bulletin 92. Sorghum for Silage and Forage.
Ohio Agricultural Experiment Station, Bulletin 21. Sorghum.
Georgia Agricultural Experiment Station, Bulletin 86. Sorghum *vs.* Corn Meal as a Source of Carbohydrates for Dairy Cattle.
Iowa Agricultural Experiment Station, Bulletin 55. Field Experiments with Sorghum.
American Breeders' Association. III. Breeding of Grain Sorghums.
Bureau of Entomology, United States Department of Agriculture, Bulletin 85. The Sorghum Midge.
South Carolina Agricultural Experiment Station, Bulletin 88. Sorghum as a Sirup Plant.
Nebraska Agricultural Experiment Station, Bulletin 77. Poisoning of Cattle by Corn or Sorghum, and Kafir Corn.
Texas Agricultural Experiment Station, Bulletin 99. Kafir Corn and Milo Maize for Fattening Cattle.
Colorado Agricultural Experiment Station, Bulletin 93. Colorado Hays and Fodder.
Texas Agricultural Experiment Station, Bulletin 13. Sorghum; Value as Feed. Effect on Soil.
Ohio Agricultural Experiment Station, Bulletin 115. Sugar Beet and Sorghum Investigations in 1899.
Delaware Agricultural Experiment Station, Bulletin 44. Sorghum in 1898.
Delaware Agricultural Experiment Station, Bulletin 27. Tests of Sorghum Varieties.
New Mexico Agricultural Experiment Station, Bulletin 33. Feeding Non-saccharine Sorghums.

INDEX

Printed in the United States of America.

Coöperation in Agriculture

By G. HAROLD POWELL

Illustrated. Cloth, 12mo, $1.50

This book deals with the general principles of coöperation. How to organize coöperative societies, how to finance them, simple organizations and constitutional documents, by-laws, and general advice as to the administration of the associations or societies are all considered. The author describes at some length the most famous organizations, such as those which are handling citrus fruits in California, the farmers' grain elevators systems, and the present coöperation in the creamery and butter business. It is, in other words, a practical guide for those who desire to organize coöperative societies and who wish to escape the usual pitfalls.

Farm Forestry

By E. G. CHEYNEY

Illustrated. Cloth, 12mo, $1.50

This book deals with the place of the wood lot or farm forest in the scheme of farming, with the planting, care, and harvesting of timber on lands, with the different species of trees that may be used, their relations or associations in a forest plantation, the rate of growth, the profits to be expected and the principal difficulties that are usually encountered. It is profusely illustrated.

Forage Crops for the South

By S. M. TRACY

Illustrated. Cloth, 12mo, $1.50

Professor Tracy has had long experience in Southern agriculture, both in application and in teaching. He was formerly Professor of Agriculture in the Mississippi Agricultural College, and now conducts a branch station or farm for the United States Department of Agriculture. He is a botanist of note and has traveled extensively in the South as a collector. His book is not only authentic, but practical. In it is contained a discussion of all kinds of plants and crops adapted to the Southern States for fodder, soiling, pasturing, and hay. The text is abundantly illustrated.

THE MACMILLAN COMPANY

Publishers 64-66 Fifth Avenue New York

Fruit Insects

By M. V. SLINGERLAND and C. R. CROSBY

Illustrated, cloth, 12mo, $1.50

This is a practical account of the principal insects in this country which attack the different kinds of fruits — trees, fruits, small fruits, cranberries, grapes, and the like. It presents a full but brief outline life history of the leading insects together with the methods of control.

Milk and Its Products

By HENRY H. WING

Professor of Dairy Husbandry in Cornell University

New Revised Edition, with new illustrations, cloth, 12mo, $1.50

The revolution in dairy practice, brought about by the introduction of the centrifugal cream separator and the Babcock test for fat, by a more definite knowledge regarding the various fermentations that so greatly influence milk, and the manufacture of its products, have demanded the publication of a book that shall give to the dairyman, and particularly to the dairy student, in simple, concise form, the principles underlying modern dairy practice. Such has been Professor Wing's purpose in this work. This is not a new edition of the author's very successful volume published under the same title many years ago; it is, in reality, an entirely new book, having been wholly reset and enlarged by the addition of new matter, both text and illustrations. The author's aim has been at all times to give the present state of knowledge as supported by the weight of evidence and the opinions of those whose authority is highest.

OF KINDRED INTEREST

The Farmer of To-morrow

BY F. I. ANDERSON *Cloth, 12mo, $1.50*

There has been a great deal of theorizing about the "back to the land" movement. It is the purpose of this book to crystallize and to make practical all of the vague generalizations which have so far been expressed on this subject. To this end the first part of Mr. Anderson's book is given over to a consideration of the land itself as a factor in the movement, primarily its economic bearing on the question. The second half takes up the soil with a detailed exposition of soil sanitation, the author confining himself, however, to only the broad principles. In presenting these two main thoughts the author touches upon such important and interesting topics as Why Europe Raises Three Bushels of Grain to Our One, Why Soils Become Unproductive, Why the Farmer of Yesterday is Rich, Why There Has Been No Increase in Acreage Productivity, and Why Irrigated Land Pays Interest on a Capitalization of Two Thousand Dollars an Acre. The book is one which should be of interest alike to those who are actively engaged in some form of agriculture and to those who are trying to solve the problem of the high cost of living.

Malaria: Cause and Control

BY WILLIAM B. HERMS *Illustrated, cloth, 8vo, $1.50*

The awakening of the general public to the necessity and possibility of the control of malaria, indicated by the incessant demand for information, makes the publication of Professor Herms's concise treatment of the subject an important and timely event. The question of malaria control is deserving of the most careful attention, particularly in these days when so much is heard of the "back to the soil" movement. For malaria is notably a disease of rural districts. Those who are familiar with the situation know very well that malaria is too often responsible for farm desertion. Professor Herms writes of the conditions attending the disease as he has found and studied them during the past few years himself, and the suggestions for control which he makes are such as he has applied and found successful.

THE MACMILLAN COMPANY

Publishers 64-66 Fifth Avenue New York

Made in the USA
Lexington, KY
02 June 2014